Guided Activities

The Practice of Social Research

TWELFTH EDITION

Earl Babbie
Chapman University

Prepared by

Theodore C. Wagenaar
Miami University

Earl Babbie
Chapman University

WADSWORTH
CENGAGE Learning™

Australia • Brazil • Japan • Korea • Mexico • Singapore • Spain • United Kingdom • United States

WADSWORTH
CENGAGE Learning™

ISBN-13: 978-0-495-59847-3
ISBN-10: 0-495-59847-X

Wadsworth
10 Davis Drive
Belmont, CA 94002-3098
USA

Cengage Learning is a leading provider of customized learning solutions with office locations around the globe, including Singapore, the United Kingdom, Australia, Mexico, Brazil, and Japan. Locate your local office at:
www.cengage.com/international

Cengage Learning products are represented in Canada by Nelson Education, Ltd.

To learn more about Wadsworth, visit **www.cengage.com/wadsworth**

Purchase any of our products at your local college store or at our preferred online store **www.ichapters.com**

Printed in Canada
1 2 3 4 5 6 7 12 11 10 09 08

Contents

Preface

Preface

This book is designed for use in conjunction with *The Practice of Social Research*, 12th edition, by Earl R. Babbie. Its purpose is to reinforce and extend your understanding of the information in the text and to provide opportunities for you to apply the material in the textbook.

Each chapter begins with objectives that focus your attention on the major points. They are specific statements of expected learning outcomes, and they will enhance your learning if you complete them. Glance through them before reading the chapter and refer to and complete them as you progress through the chapter. An outline and a summary of the chapter follow. Reading the summary before and after reading the chapter will give you both a preview and a review of its contents.

A list of the key terms follows the summary. Mark each in the text and try to restate the definition in the text. Several of the definitions for these terms are presented in the matching exercises to give you some practice. The answers and page number references are contained in Appendix 2 for your convenience. The true-false and review questions test both knowledge and the ability to apply the concepts and principles presented in the text. Be sure to complete them and check your answers with the correct answers in Appendix 2. Page numbers are provided in the event you wish to determine why you answered a particular question incorrectly. Completing these questions will enhance your performance on exams.

The discussion questions encourage you to address several of the major issues raised in the chapter. Writing out the answers in your own words will strengthen your understanding and application of these issues.

The best way to learn research methods is through actual practice. Hence, we have included several short exercises for each chapter. Several of them involve you in analyzing data from the General Social Surveys by using SPSS, MicroCase, or another data analysis program. Before starting work on an exercise, we suggest you read through the entire exercise first so that you will know everything that is required. Reread those portions of the chapter that deal with the topics covered in the exercise. Then you should be ready to start work. In suggesting so many different exercises, it was our intention to provide you and your instructor with options. The exercises are all limited enough in scope that you will be able to complete many of them during the course. But we certainly did not intend for you to do them all. Make sure your instructor knows that. Instructors are reminded that suggested answers for some of the exercises are presented in the *Instructor's Manual*. Some of the Internet exercises provided are modified versions of those provided in previous editions of the *Instructor's Manual* (by Margaret Jendrek).

Comments and suggestions are welcome. Please address comments to Theodore C. Wagenaar, Miami University, Department of Sociology and Gerontology, Oxford, OH 45056. Email: wagenatc@muohio.edu.

Part 1

An Introduction to Inquiry

Chapter 1

Human Inquiry and Science

OBJECTIVES

1. Compare knowing things through agreement versus through experience.

2. Identify the two criteria needed for scientists to accept the reality of something they have not personally experienced.

3. Differentiate epistemology from methodology.

4. Define and illustrate causal reasoning and probabilistic reasoning.

5. Differentiate the scientific approach from the ordinary human inquiry approach to causal and probabilistic reasoning.

6. Differentiate prediction from understanding.

7. Describe the role of tradition and authority as sources of secondhand knowledge.

8. Define and illustrate each of the following errors in inquiry: inaccurate observations, overgeneralization, selective observation, and illogical reasoning.

9. Show how a scientific approach provides safeguards against each one of these errors.

10. Describe premodern, modern, and postmodern views of reality.

11. Describe the three major aspects of the overall scientific enterprise.

12. Define theory and indicate how it differs from philosophy or belief.

13. Give three examples of social regularities.

14. Identify and assess the three objections that are sometimes raised in regard to social regularities regularities.

15. Define aggregate and present a rationale for why social scientists examine aggregates.

16. Give four examples of variables and their respective attributes.

17. Differentiate independent and dependent variables by definition and example, and show how they contribute to understanding causality.

18. Compare the three approaches for doing research that vary by purpose: exploratory, descriptive, and explanatory.

19. Compare idiographic and nomothetic explanations.

20. Compare inductive and deductive theory.

21. Compare quantitative and qualitative data.

22. Differentiate pure from applied research.

OUTLINE

1. Looking for reality
   a. Ordinary human inquiry
   b. Tradition
   c. Authority
   d. Errors in inquiry and some solutions
   e. What's really real?

2. The foundations of social science
   a. Theory, not philosophy or belief
   b. Social regularities
   c. Aggregates, not individuals
   d. A variable language
   e. The purposes of research
   f. The ethics of human inquiry
3. Some dialectics of social science
   a. Idiographic and nomothetic explanation
   b. Inductive and deductive theory
   c. Qualitative and quantitative data
   d. Pure and applied research

SUMMARY

People know what they know primarily through two processes. First, people know things through agreement, the things they consider to be real because they have been told they are real. Believing what we are told yields most of what we know. Second, people also know things through direct experience, primarily through observation. People have long struggled with determining what is real.

Science provides a way of looking at both agreement reality and experiential reality. The scientific approach to both realities demands that two criteria be met: an assertion must have both logical and empirical support. Hence science is a special form of human inquiry, the result of the human desire to predict future events and to understand patterns of cause and effect. The textbook will examine social science methodology, which could simply be called the science of finding out. Methodology is a sub-field of epistemology, the science of knowing.

Recognizing that some factors are caused by other factors lies at the root of ordinary human inquiry. That is, we use causal reasoning. Such patterns of cause and effect are probabilistic. That is, the cause usually but not always brings about the effect. Understanding ordinary human inquiry also means distinguishing prediction from understanding. While it is possible to predict something without necessarily understanding why it occurs, understanding usually yields improved prediction abilities. In short, ordinary human inquiry involves both what and why questions.

Tradition and authority comprise two important sources of agreement reality, our secondhand knowledge. Using tradition, we accept an inherited body of information and understanding. Tradition helps us avoid having to learn everything from scratch. Authority derives from the status of the transmitter of the knowledge. We tend to believe those with more credibility. But both authority and tradition can also hinder human inquiry.

Ordinary human inquiry is prone to several errors. First, casual observations are frequently inaccurate. Science helps alleviate this error by mandating conscious observation. Second, people frequently overgeneralize on the basis of a few limited observations. Scientists protect themselves against overgeneralization by employing large random samples and by replicating studies. Third, people observe selectively, by paying attention to events that match a prior conclusion and ignoring those that do not. The scientific approach helps protect against this error by specifying in advance the number and types of observations to be made and by having several scientists investigate the same phenomenon. A fourth error is illogical reasoning, such as using an exception to prove a rule. Science helps here by providing systems of logic. In comparison to casual inquiry, scientific inquiry is a conscious activity and is more careful than our casual inquiry.

Most people give little thought to the nature of reality, something philosophers refer to as "naive realism." The nature of "reality" is more complex than we tend to assume, and can be approached through three views of reality. The premodern view of reality played a key role in earlier civilizations and assumed that things really were the way humans saw them. The modern view considers differing views of reality as legitimate; there is no one best view of what is assumed to exist. The postmodern view holds that what is "real" simply reflects our points of view and that there is actually no "objective" reality to be observed. The postmodern view highlights the difficulty scientists have in observing and explaining what is "real" because scientists' personal orientations can affect what they observe and how they explain what they observe.

Science is based on logic and observation. The scientific enterprise involves theory, data collection, and data analysis. Theory meets the scientific criterion of logic by describing the logical relationships that exist among variables; it contributes to our understanding of what is observed. Data collection meets the scientific criterion of observation. Data analysis uncovers patterns in what is observed and helps compare what is logically expected with what is actually observed. Theory addresses what is, not what should be. Hence scientists cannot settle debates on values. In short, social science can only help in knowing what is and why.

Social scientific theory and research help uncover the social regularities under-lying social life. Some people argue that social regularities are sometimes trivial; scientists point out that commonsense understandings are often incorrect. Others argue that contradictory cases negate social regularities; scientists point out that social regularities represent probabilistic patterns that need not apply to every situation. Still others argue that individuals could upset a social regularity if they so desired; scientists point out that such action is possible but unlikely given the social norms that regulate social behavior. In short, social scientists analyze aggregates instead of individuals by examining the relationships between variables. They do so by examining the distributions of people across the attributes, or values, of variables that reflect causes (independent variables) and effects (dependent variables). Understanding causal connections remains paramount.

Social scientists pursue three purposes in doing research. Exploratory research helps map out a topic for further research. Descriptive research helps describe some aspect of social reality. Explanatory research helps provide reasons for social phenomena. Ethical issues enter in all three types.

Four broad and connected distinctions underlie many of the variations of social research. The first involves two types of causal reasoning. Idiographic explanations address all or most of the unique factors that help explain one particular case. Nomothetic explanations, on the other hand, address the major factors that help explain something across many cases. Nomothetic explanations seek explanations with fewer causal variables and seek to explain the causal connection in a variety of cases.

The second distinction involves two types of theory. Inductive reasoning moves from particular observations to the discovery of a pattern underlying the particular observations. Deductive reasoning, on the other hand, moves from a pattern that might be logically or theoretically expected to observations that confirm or disconfirm the expected pattern.

The third distinction involves the difference between numerical and non-numerical data. Quantitative analysis helps make our observations more explicit and affords statistical analyses. Qualitative analysis provide greater detail and provide a greater richness of meaning than does quantitative analysis.

The fourth distinction involves two motivations for scientists. Pure research seeks knowledge for knowledge's sake alone, while applied research seeks to apply knowledge to everyday problems in order to help address them.

TERMS

1. aggregates
2. attributes
3. applied research
4. authority
5. causal reasoning
6. causation
7. data collection
8. data analysis
9. deductive theory
10. dependent variable
11. descriptive
12. empirical
13. epistemology
21. logical
22. methodology
23. modern view of reality
24. nomothetic explanation
25. overgeneralization
26. postmodern view of reality
27. prediction
28. premodern view of reality
29. probabilistic reasoning
30. pure research
31. qualitative data
32. quantitative data
33. relationship

14.	explanatory	34.	replication
15.	exploratory	35.	selective observation
16.	idiographic explanation	36.	social regularities
17.	illogical reasoning	37.	theory
18.	inaccurate observation	38.	tradition
19.	independent variable	39.	understanding
20.	inductive theory	40.	variables

MATCHING

(A reminder: the answers to matching, true-false, and review questions are in Appendix 2)

1. The science of finding out.

2. Explanation that seeks to identify the major factors that affect something across many cases.

3. A source of knowledge that is obtained from experts.

4. The assumption that a few similar events are evidence of a general pattern.

5. Interpreting events to fit a general pattern that a researcher believes to be true.

6. Logical and persistent patterns in social life.

7. Logical groupings of attributes.

8. An association that links variables.

9. Collective actions and situations of many individuals.

10. The cause of a particular dependent variable.

TRUE-FALSE QUESTIONS

T F 1. Epistemology is the science of knowing and methodology is the science of finding out.

T F 2. Overgeneralization is more likely to lead to selective observation than selective observation is to lead to overgeneralization.

T F 3. The modern view of reality states that what is "real" depends on our points of view.

T F 4. One of the unique advantages of science is that it can settle debates on values.

T F 5. The argument that there are always exceptions to patterns helps dispel the notion of social regularities.

T F 6. Attributes are composed of variables.

T F 7. Nomothetic explanations seek to explain a class of situations or events rather than a single one.

T F 8. Inductive reasoning involves going from the specific to the general.

T F 9. While nomothetic explanations are very useful in social scientific research, idiographic explanations are of little use.

T F 10. Pure research pursues knowledge for knowledge's sake, with no concern for how it might be used.

MULTIPLE-CHOICE QUESTIONS

1. Most of what we know is a matter of
    a. personal experience.
    b. a result of scientific discovery.
    c. empirical evidence.
    d. agreement reality.
    e. logical support.

2. Which one of the following is the best example of agreement-based knowledge?
    a. you meet 15 Graskinos and conclude that they are all prejudiced
    b. you develop a scale to measure love and give it to 20 people
    c. you accept prejudice as wrong because that is what the Bible says
    d. you do not get right next to people when you talk to them because it violates a norm
    e. you run an experiment to test the effect of crowding

3. The two foundations of science are
    a. tradition and observation.
    b. observation and logic.
    c. logic and theory.
    d. theory and observation.
    e. logic and generalization.

4. When you hear the 4:27 train blow its whistle as you walk home from school every day, you can expect within minutes to smell Ms. Stockland's cooking. This is an example of
    a. understanding without prediction.
    b. understanding with prediction.
    c. prediction with understanding.
    d. prediction without understanding.
    e. scientific reasoning.

5. "But my professor said that no significant differences exist between men and women regarding intelligence." What source of understanding does this example reflect?
    a. personal experience
    b. tradition
    c. authority
    d. public opinion
    e. science

6. A student meets two fraternity men at a party who talk about all the partying they do. She then concludes that all fraternity men party all the time. What error in understanding does this example reflect?
    a. inaccurate observation
    b. overgeneralization
    c. selective observation
    d. illogical reasoning
    e. overemphasis on authority

7. In your discussion of measurement with a friend, she argues that what you are trying to measure does not exist and that your own point of view will determine what you perceive in your measuring process. She has taken which view of reality?
    a. correct
    b. premodern
    c. modern
    d. postmodern
    e. scientific

8. You have just had a wonderful streak of great luck in your methods class: you've gotten A's on the last three tests and computer assignment. You have a research project due on the last day of class and you just know that you are going to flunk it. After all, something has to happen to break up this streak of good luck. You have fallen prey to the error of
    a. illogical reasoning.
    b. inaccurate observation.
    c. selective observation.
    d. over-emphasis on tradition.
    e. overgeneralization.

9. Criminal justice student Vierling has noticed in her internship that adolescents who have experienced difficulties in school are more likely to become juvenile delinquents. In her process of going from specific observations to the discovery of a more general pattern, she has employed
    a. social regularity thinking.
    b. illogical reasoning.
    c. deductive reasoning.
    d. inductive reasoning.
    e. nomothetic inquiry.

10. Science
    a. deals with what should be and *not* with what is.
    b. can settle debates on value.
    c. is exclusively descriptive.
    d. has to do with disproving philosophical beliefs.
    e. has to do with how things are and why.

11. Scientists do not study individuals per se, but instead study social patterns reflecting
    a. aggregates.
    b. collectivities.
    c. theories.
    d. attributes.
    e. norms.

12. Professor Flemming examined the following categories of marital status: married, never married, widowed, separated, and divorced. These categories are known as
    a. variables.
    b. attributes.
    c. variable categories.
    d. units of analysis.
    e. theoretical elements.

13. When social scientists study variables in explanatory research, they focus on
    a. attributes.
    b. groups.
    c. people.
    d. characteristics.
    e. relationships.

14. Senator Josephson researched the effects of political orientation on attitudes toward abortion. "Political orientation" is an example of
    a. an attribute.
    b. an independent variable.
    c. an aggregate.
    d. a unit of analysis.
    e. a dependent variable.

15. Dunn was interested in learning more about the negotiation processes couples use when they argue. He was interested in this for its own sake and had no plans to use the results to help improve couples' lives. What type of research is reflected in this example?
    a. funded
    b. applied
    c. pure
    d. ethical
    e. idiographic

16. Bernardo argues that nothing is really real and that what people see is a product of their position in society. Which perspective is reflected?
    a. modern
    b. premodern
    c. postmodern
    d. anti-modern
    e. pesudomodern

17. Igor studied students at his community college and noticed that a slim majority carried no books to class. Then he saw a new set of people in the distance and concluded before seeing them that they would definitely be carrying books to class because someone should do that, especially since he had just seen students without books. Which error in inquiry is reflected?
    a. illogical reasoning
    b. overgeneralization
    c. selective observation
    d. inaccurate observation
    e. scapegoat

18. Makita studied the literature on binge drinking and discovered a pattern between gender and binge drinking that might be logically and theoretically expected. She then did her study of college students and drinking. Which approach is reflected?
    a. deductive
    b. transductive
    c. applied
    d. ethical
    e. inductive

19. Juanita decided to study the differences among students who choose different majors. Everyone in her study, however, has chosen sociology as a major. Which one of the following can be said about her concept of choice of major?
    a. she needs to study more people
    b. she can go ahead and do her study relating the two variables
    c. she really has only one concept
    d. one of her variables isn't really a variable because it has only one attribute
    e. it is far too idiographic to be useful

20. Van says that his 48-year old Catholic uncle will marry his 23-year old atheist girlfriend next year in order to prove that the social regularity that people tend to marry people like themselves is not accurate. What is your best response as a methodology student?
    a. the objection that social regularities could be upset through the conscious will of the actors is not a serious challenge to social science
    b. Van's uncle is simply an exception to the rule
    c. this regularity is too trivial to be meaningful
    d. we need to develop a measure for "marry people like themselves"
    e. Van fell victim to the error of overgeneralization

DISCUSSION QUESTIONS

1. How do tradition and authority hinder inquiry? How do they support it? Give examples.

2. Present both sides of the argument that the scientific approach is the best approach to understanding.

3. Show how the three views of reality reflect the intellectual climate of the time during which they developed.

4. Compare deductive reasoning and inductive reasoning by definition and example.

EXERCISE 1.1

Name_____

Each of us is confronted with decisions in our everyday lives that require us to gather and assess information on the different alternatives at hand and then make a decision. Examples of such decisions include the decision to attend college, buy a car or some other item, strike up a friendship with Person A or B, select a particular course, or take a trip to Point X or Y. You may have made an error in such decisions because your information was flawed by one or more of the errors of human inquiry that Babbie describes, or the decision may have been correct but for some of the wrong reasons. Recall and describe a decision you have made that may have been flawed to some extent because information was based on one or more of the errors of human inquiry.

1. Describe the decision.

2. Identify which of the errors of human inquiry as described by Babbie were involved in this decision. Explain how each error was committed.

3. Explain how a scientific approach would have helped reduce the effect of the error(s) of human inquiry in this decision.

EXERCISE 1.2   Name_____

List four social (not physical) regularities that are evident in day-to-day life.
These could involve regularities in family life, friendship, large groups, work
experiences, classroom activities, etc. Be sure you are describing social
regularities. See the text for examples, but do not use these examples. Again,
avoid physical examples.

EXERCISE 1.3   Name_____

Variables and their attributes (or values) are at the heart of examining relationships in the social sciences. Key to formulating such relationships are independent and dependent variables. Select two relationships, each involving both an independent variable and a dependent variable. State each of the two relationships in a separate sentence. Each sentence should contain just two variables—an independent variable and a dependent variable—and should contain words which specify how the two variables are related to each other. The independent variable should be stated first in the sentence. You should then identify which is the independent variable and which is the dependent variable in each statement of a relationship. Explain why you expect there to be a relationship between the two variables. Then identify clearly the attributes of each variable.

Review the examples in the text if you have difficulty with this exercise. Be sure to use sociological variables and avoid those used in the text or by your instructor.

A few examples of incorrect and correct examples follow.

Incorrect: "Men like Fords better than Volvos." This is incorrect because there is only one variable, type of car liked. Men in this sentence is a constant, not a variable. *Correct*: "Men like cars more than women do." The independent variable is gender (attributes: male, female) and the dependent variable is liking cars (attributes: like much, like some, like little, etc.).

Incorrect: "High school graduates and college graduates hold jobs." This is incorrect because it really specifies a constant (holding jobs) that is associated with both attributes of the variable amount of education. *Correct*: "College graduates hold higher paying jobs than do high school graduates." The independent variable is amount of education (attributes: high school graduate, college graduate) and the dependent variable is amount of income (attributes: less than $5,000, $5,001-$10,000, etc.).

Incorrect: "Intelligence is related to grades." This statement does have two variables, but it is incorrect because the nature of the relationship is not specified. *Correct:* "The higher people's intelligence, the higher will be their grades." The independent variable is amount of intelligence (attributes: less than 70, 71-90, etc.) and the dependent variable is grades (attributes: A, B, etc.).

1. For your first relationship, identify the independent and dependent variables, explain why you expect there to be a relationship, and describe the attributes of each. Remember to use sociological variables.
   a. The relationship:

   b. The independent variable:

   c. The dependent variable:

   d. Why do you expect a relationship:

   e. The attributes of the independent variable:

   f. The attributes of the dependent variable:

(Continued)

2. For your second relationship, identify the independent and dependent variables, explain why you expect there to be a relationship, and describe the attributes of each. Remember to use sociological variables and do not use either of the two variables you used in #1.

    a. The relationship:

    b. The independent variable:

    c. The dependent variable:

    d. Why do you expect a relationship:

    e. The attributes of the independent variable:

    f. The attributes of the dependent variable:

EXERCISE 1.4

Name_____

Visit the Web site of the American Sociological Association (www.asanet.org). Click "Sociologists" and then click "Data Resources Available." Review the publicly available data sets described and select one (other than the General Social Survey) that interests you.

1.   Identify the data set you chose.

2.   Describe the aggregate(s) the data set examines.

3.   List three social regularities that could be examined with the data set.

4.   Identify a dependent variable that is or might be included in the data set and note two independent variables that may be connected to that dependent variable.

EXERCISE 1.5

Name_____

Examine the variables reported in the General Social Survey, Appendix 1 in this volume. Identify five variables of interest, but pick those with relatively few attributes. Run a frequency distribution of these variables, using SPSS or another data analysis program as indicated by your instructor.

1. For each variable:
   1. list the variable
   2. describe the variable—see the label in Appendix 1
   3. list the attributes
   4. list the percentage in each attribute (use the valid percent column, the one with the missing values omitted from the percentage calculations)

2. Select three of your variables and briefly summarize the distribution of cases across the attributes.

EXERCISE 1.6

Name_____

Babbie discusses social scientists' desire to identify social regularities. One regularity is that men earn more than women do. Use SPSS or another data analysis program as indicated by your instructor to test this observation by running a crosstabulation of income (INCOME3) by gender (SEX) (see Appendix 1). Be sure to put SEX on top (columns) and INCOME3 on the side (rows), and request column percentages.

1.   Show your table below.

2.   Interpret the results.

ADDITIONAL INTERNET EXERCISES

1. See the discussion of the scientific method at http://www.scientificmethod.com/ and read the following three sections: 1) scientific methods vs. the scientific method, 2) the 11 stages and 3 supporting ingredients of the SM-14 formula, and 3) practical help with everyday problems and decisions. Address these questions: 1) what are the stages in the scientific method?; 2) why is the scientific method used?; and 3) what does the scientific method enable researchers to conclude?

2. http://www.scientificmethod.com/Visit http://www.aic.gov.au/conferences/outlook4/Borowski.html and read the article, "The Dangers of Strong Causal Reasoning in Policy and Practice: The Case of Juvenile Crime and Corrections." We usually think of strong causal reasoning as a good thing. Why does Borowski argue that strong causal reasoning may be dangerous? What are his conclusions about the roles of deductive and inductive reasoning in research on juvenile crime and corrections?

3. Visit the Social Research Update site at http://www.soc.surrey.ac.uk/sru/ and select an article/site that interests you. Show its relevance for research methods.

Chapter 2

Paradigms, Theory, and Research

OBJECTIVES

1. List the three functions of theory in research.

2. Define paradigm and describe their role in science.

3. Note the two benefits of knowing that we operate within a paradigm.

4. Differentiate macrotheory from microtheory.

5. Provide synopses for each of the following paradigms: early positivism, Social Darwinism, conflict, symbolic interactionism, ethnomethodology, structural functionalism, feminist paradigms, critical race theory, and postmodernism.

6. Show the role of theory, operationalization, and observation in the traditional model of science.

7. Define hypothesis testing.

8. Differentiate inductive logic from deductive logic by definition and example.

9. Outline the steps in deductive theory construction.

10. Show how ethical issues enter our thinking about theories and paradigms.

OUTLINE

1. Some social science paradigms.
   a. Macrotheory and microtheory
   b. Early positivism
   c. Social Darwinism
   d. Conflict paradigm
   e. Symbolic interactionism
   f. Ethnomethodology
   g. Structural functionalism
   h. Feminist paradigms
   i. Critical race theory
   j. Rational objectivity reconsidered

2. Elements of social theory

3. Two logical systems revisited
   a. The traditional model of science
   b. Deduction and induction compared

4. Deductive theory construction
   a. Getting started
   b. Constructing your theory
   c. An example of deductive theory

5. Inductive theory construction
   a. An example of inductive theory

6. The links between theory and research

7. The importance of theory in the "real world"

8. Research ethics and theory

SUMMARY

Chapter 1 showed how scientific inquiry is a product of both logic and observation. Chapter 2 formalizes that analysis by examining the respective roles of theory and research. The close link between the two is again demonstrated. It is risky to base research only on the observation of patterns. Social scientists use theory to provide logical explanations for observed patterns. Theory has three functions for social research. First, it helps prevent us from being taken in by flukes. Second, theories help us to make sense out of observed patterns and enable us to suggest other possibilities. Third, theories can shape and direct research efforts by suggesting likely discoveries through empirical observation. However, not all social science research is tightly connected with theory.

Social scientists have used various frames of reference, known as paradigms, in their search for meaning. Whereas theories seek to explain, paradigms provide ways of looking–they provide logical frameworks within which theories are created. Paradigm usage helps us to better understand those operating under different paradigms. Paradigms also encourage us to move beyond our own favorite ones to others that shed a different light on what we wish to understand. The paradigms scientists employ help determine which theories seem the most useful. Paradigms are often difficult to recognize as such because they are so implicit, assumed, and taken for granted.

Paradigms often gain or lose popularity, but are seldom discarded completely. Different paradigms make different assumptions about the nature of social reality. Paradigms range from the macro level to the micro level. Macrotheory examines society at large or at least large portions of it, while micro-theory examines social life at the levels of individuals and small groups. Mesotheory is sometimes used to refer to an intermediate level between the two.

Many paradigms exist. Early positivism was founded by Comte, who was the first to suggest that society could be studied logically and rationally. His emphasis on positivism laid the foundation for the development of social science. Social Darwinism is found in the works of Spencer and reflects the idea that societies continue to improve as they adapt to changes in their surroundings. The conflict paradigm suggests that conflict underlies much of social interaction. Conflict theorists have explored macro-level conflict (such as Marx's analysis of conflict among different economic classes) and micro-level conflict (such as Simmel's analysis of conflict within groups). Conflict theory illustrates how paradigms shape the kinds of observations we are likely to make, the facts we discover, and the conclusions we draw from those facts. Paradigms also influence our choice of relevant concepts.

Symbolic interactionists examine interactions among individuals and focus on how people gain a sense of self based on their interactions with others. They stress the use of language and symbols in their analyses. Ethnomethodologists believe that people continually create social structure through their actions and interactions. By continually trying to make sense out of their lives, people create their own social realities. Violating social norms often helps us determine people's expectations and how they make sense of their world. Structural functionalism is based on the assumption that social systems are made up of various parts (structures), each of which contributes to the functioning of the whole social system.

The feminist paradigm shows how gender differences relate to the rest of social organization. The analysis of the oppression of women in various societies has contributed to our understanding of oppression in general. The feminist paradigm questions the predominant views of social reality because such views have generally been developed by men. Feminist paradigms often highlight the limitations in how aspects of social life are typically examined and understood. Feminist standpoint theory refers tot he fact that women have knowledge about their status and experience that is not available to men.

Critical race theory highlights how race affects the questions asked, the analyses offered, and the conclusions drawn about social phenomena. It emerged in the 1970s with a codification of a paradigm based on race awareness and a commitment to racial justice. But it has its origins in the work of W. E. B. DuBois who noted that African Americans experience a "dual consciousness" as simultaneously American and as black people. Much of the contemporary scholarship in critical race theory examines the role of race in politics and government.

Some scholars question the assumption of rationality underlying the growth in science and in the rise of bureaucratic structures. People do not always behave rationally. However, it is possible to study even nonrational behavior in a scientific manner. This approach also challenges the idea that scientists can really be as objective as the ideal image of science assumes. Postmodernists have questioned the belief that there is a logically-ordered, objective reality and that scientists can study it objectively. All human experiences, including those of scientists, are inescapably subjective. A parallel theory, critical realism, suggests that people define reality as that which can be seen to have outcomes.

The paradigms described above differ somewhat from theory. Paradigms are general frameworks for viewing social reality and are grounded in various assumptions about the nature of reality. A theory is a systematic set of interrelated statements that seeks to explain some aspect of social life. Think of paradigms as more general and theories as more specific. A paradigm provides a way of looking and a theory strives to explain what we see.

Social scientists employ several elements in constructing theories. Observation refers to seeing, hearing, and touching. A fact means some phenomenon that has been observed, and a law reflects a universal generalization about a class of facts. There are no laws in social science that parallel the laws in the natural sciences. A theory is a systematic explanation for the observations that relate to a particular aspect of social life. Concepts are the basic building blocks of theory and are abstract elements representing classes of phenomena with the field of study.

A variable includes a set of attributes. Axioms or postulates are basic assertions assumed to be true; they help ground a theory. Propositions are conclusions drawn about the relationships among concepts; they are derived from the axiomatic groundwork. Hypotheses are specified expectations about empirical reality; they are derived from propositions.

There are three main elements in the traditional model of science: theory, operationalization, and observation. Theory provides a backdrop for clarifying possible relationships among variables. Operationalization enables the measurement of variables so that their relationships can be empirically examined. Observation involves the actual collection of data to test the hypothesized relationships. This entire strategy is also known as hypothesis testing. Theory construction proceeds through deductive logic and/or inductive logic. With deductive logic, scientists progress from general principles and theories to specific cases. With inductive logic, they proceed from particular cases to more general theories. In reality, theory and research interact through a mix of both deduction and induction.

Deductive theory construction follows these major steps: (1) specify the topic of interest, (2) specify the range of phenomena the theory addresses, (3) identify and specify the major concepts and variables, (4) find out what is known about the relationships among the variables, and (5) reason logically from the propositions obtained in the preceding steps to the specific topic under analysis. The deductive approach underscores the two basic elements in science: logical integrity and empirical verification.

In contrast to the deductive approach, the inductive approach to theory construction strives to uncover patterns based on actual observation; this approach is also known as grounded theory. This approach is seen most often in field research, although it can be used in other designs as well. Grounded theorists argue that the inductive approach is superior to deductive theorizing because the social scientist does not enter the field with preconceived theoretical constraints. The deductive and inductive approaches are two idealized logical models for linking theory and research. However, social scientists employ many variations on these two ideal models. Some studies, for example, display less connection between theory and research than do others. Some people feel that theoretical and practical issues do not overlap. However, theory often provides the insight needed to both understand society and to change it.

Ethical issues enter our thinking about theory and paradigms. Adopting a particular theoretical orientation in order to strengthen a particular conclusion would be considered unethical generally. But when researchers seek social change, they often employ theoretical orientations relevant for that intention. The potential biasing effect of this approach is mitigated by two factors. First, scientific observation and analysis reduce the propensity for simply seeing what we expect to see. Second, research is a collective effort and peer review would help uncover biased research.

TERMS

1. axioms
2. concepts
3. conflict paradigm
4. critical race theory
5. deductive reasoning
6. ethnomethodology
7. fact
8. feminist paradigms
9. grounded theory
10. hypothesis
11. hypothesis testing
12. inductive reasoning
13. law
14. macrotheory
15. microtheory
16. null hypothesis
17. observation
18. operational definition
19. operationalization
20. paradigm
21. proposition
22. social Darwinism
23. structural functionalism
24. symbolic interactionism
25. theory
26. variables

MATCHING

_____ 1. The specification of the steps, procedures, or operations followed to actually measure variables.

_____ 2. Reasoning from particular instances to generalizations.

_____ 3. A paradigm that examines interaction among individuals and how people gain a sense of self based on their interactions with others.

_____ 4. This paradigm shows how gender differences relate to the rest of social organization.

_____ 5. This paradigm uses "methodology of the people" to examine how people continually create social structure through their actions and interactions.

_____ 6. Reasoning from general principles and theories to specific cases.

_____
_____ 7. A general framework or viewpoint that organizes our view of something.

_____
_____ 8. The paradigm that focuses on how the components of society are interrelated.

_____ 9. The paradigm that focuses on how race affects our understanding of social phenomena.

_____ 10. The paradigm that is based on Marx's analysis of economic attempts to dominate others and to avoid being dominated.

TRUE-FALSE QUESTIONS

T F 1. Whereas theories seek to explain, paradigms provides ways of looking.

T F 2. Feminist paradigms look only at the oppression of women.

T   F   3. Karl Marx coined the term *sociologie* (sociology).

T   F   4. Breaking the rules is a technique employed by ethnomethodologists.

T   F   5. The traditional model of science begins with operationalization.

T   F   6. The traditional model of science employs deductive logic.

T   F   7. Identifying your major concepts and variables is the first step in constructing your theory.

T   F   8. Inductive theory is commonly done through observation research.

T   F   9. Regardless of the theory used, there is a close connection between theory and research.

T   F   10. Experimental research is commonly use to conduct inductive theory construction.

REVIEW QUESTIONS

1. Professor Peters examined the gender makeup of politicians and then made conclusions about how the political institution reflects the other institutions of society. Which paradigm was she using?
    a.   Ethnomethodology
    b.   post-positivism
    c.   role theory
    d.   feminist
    e.   exchange

2. The three main elements of the traditional model of science are
    a.   theory, operationalization, observation.
    b.   operationalization, hypothesis testing, theory.
    c.   observation, experimentation, operationalization.
    d.   theory, observation, hypothesis testing.
    e.   experimentation, hypothesis testing, theory.

3. Which of the following is the **best** example of a hypothesis?
    a. The greater the level of education, the greater the tolerance for alternative lifestyles.
    b. Socialization in childhood has a significant impact on adolescent gender-role identity.
    c. There are more female than male college students.
    d. Religiosity equals frequency of church attendance and praying.
    e. Actions are based on perceived costs and rewards.

4. Which of the following is the *best* example of macrotheory?
    a. a study of play interactions among children
    b. a study of arguments among friends
    c. a study of cheating among students
    d. a study of the role of corporate mergers in the global economy
    e. a study of tax audits conducted by the IRS

5. Professor Pellson noticed that the students who sat in the front row in her classes usually got A's. After observing this pattern over several semesters, she decided that students who sit in the front row are generally good students. This is an example of
    a. deductive reasoning.
    b. inductive reasoning.
    c. probabilistic reasoning.
    d. causal reasoning.
    e. deterministic reasoning.

6. Babbie reviews the traditional model for portraying science. This model shows that science generally begins with
    a. theories.
    b. hypotheses.
    c. observations.
    d. empirical conclusions.
    e. measurement.

7. Whose emphasis on positivism laid the foundation for the development of social science?
    a. Weber
    b. Darwin
    c. Comte
    d. Garfinkel
    e. Mead

8. Professor Havidan studied a prenatal class for expectant parents in his attempt to understand how expectant parents prepare for the role of parenthood and how they gain a sense of themselves as parents. He focused on the language and symbols expectant parents use to describe their new roles. Which paradigm is he using?
    a. role theory
    b. feminist
    c. ethnomethodology
    d. structural functionalism
    e. symbolic interactionism

9. Deductive theory construction generally begins with
    a. specifying topic of interest.
    b. specifying the range of phenomena the theory addresses.
    c. identifying and specifying the major concepts and variables.
    d. finding out what is known about the hypothesized relationships.
    e. reasoning logically from the propositions obtained to the specific topic under analysis.

10. The inductive approach to theory construction is seen most often in
    a. survey research.
    b. experimental research.
    c. field research.
    d. use of available data.
    e. content analysis.

11. The paradigm that accounts for the impact of economic conditions on family structures is
    a. symbolic interactionism.
    b. structural functionalism.
    c. positivism.
    d. conflict.
    e. exchange.

12. Which of the following is *not* a step in deductive theory construction? Or are they all steps?
    a. specify the topic
    b. identify the major concepts and variables
    c. identify propositions about the relationships among those variables
    d. reason logically from those propositions to the specific topic one is examining
    c. all arc stcps

13. Professor Todd observes playground interaction for a month to develop a theory that accounts for the problems observed. This is an example of
    a. deductive theorizing.
    b. grounded theory.
    c. conceptualization.
    d. ex post facto hypothesizing.
    e. paradigm estimation.

14. Indira wants to study how people make sense of the delays they experience in their daily lives. She asked people to keep diaries about such delays and how they responded to them, particularly how they made sense of them. She also encouraged them to sometimes violate others' expectations in terms of responding to delays. Which paradigm is she using?
    a. ethnomethodology
    b. symbolic interactionism
    c. structural functionalism
    d. conflict
    e. feminist

15. Roberto wants to study how newly engaged couples communicate with each other. He is particularly interested in how each person defines such symbols as the engagement ring, and how each person communicates the relationship to others. Which paradigm would be best?
    a. ethnomethodology
    b. symbolic interactionism
    c. structural functionalism
    d. conflict
    e. feminist

16. Who first coined the term "sociology?"
    a. Durkheim
    b. Marx
    c. Comte
    d. Mead
    e. Cooley

17. Who was an early theorist concerned with how individuals interacted with one another?
    a. Durkheim
    b. Simmel
    c. Mead
    d. Spencer
    e. Parsons

18. Frankie assumed that kids like to be respected by other kids. He then developed a specific testable expectation that boys experience more pressures for delinquency than do girls. This expectation is known as a/an:
    a. hypothesis
    b. concept
    c. variable
    d. proposition
    e. conclusion

19. Freda developed a theory and an hypothesis about adjustment to retirement as related to gender. She constructed measures for adjustment to retirement. The next step for Freda according to the traditional model of science is:
    a. to consider the ethics of the study
    b. redevelop the theory
    c. analyze the data
    d. observation (gathering data)
    e. construct a sample

20. Which one of the following is ***not*** a function of theories?
    a. they prevent our being taken in by flukes
    b. they make sense of observed patterns to suggest other possibilities
    c. they shape and direct research efforts
    d. they help identify the more appropriate ways to view the world
    e. all of these are functions of theory

DISCUSSION QUESTIONS

1. Babbie discusses several paradigms: early positivism, conflict, symbolic interactionism, ethnomethodology, structural functionalism, the feminist paradigm, and critical race theory. Briefly summarize each of these, as well as his points on rational objectivity reconsidered. Then select one topic of interest and show how three of the paradigms would examine that topic.

2. Figure 2-2 in the text portrays the traditional image of science. Describe that model, being certain to discuss the role of the three main elements: theory, operationalization, and observation.

3. Describe how the elements of theory construction fit together. Be specific.

4. Contrast deductive and inductive logic. Describe specific research issues that are appropriate for each and explain why.

5. Show how theory and research are closely linked. Indicate what each contributes to the other.

EXERCISE 2.1

Name _____

A major focus of this chapter is the difference between deductive and inductive logic. Your assignment is to develop studies using each approach.

DEDUCTIVE LOGIC

1. Identify a topic of interest. You might select religiosity, feminism, or occupational success if you cannot identify a topic.

2. Explain what your theory will address. It might be factors promoting religiosity, why some people are more feminist than others, or why some people are more successful in their occupations than others.

3. Specify the range of phenomena your theory addresses. All people? Women only? Americans only?

(continued)

4. Identify and specify your major concepts and variables. Don't just list your major concepts and variables, but actually develop and state your theory. Your theory does not have to be very complex. Keep in mind that a theory is an explanation of the causes of some phenomenon. Your theory may have one or more statements in it.

5. Derive at least one specific testable hypothesis, such as: younger females are more likely to be feminist than older females. Be sure your hypothesis reflects a specific relationship between two variables.

INDUCTIVE LOGIC

1. Select one of the following "observations" or select one of your own. If you select one of your own, make sure it is very specific.
   a. women like coed residence halls more than men do
   b. students prefer male instructors over female instructors
   c. older people are more likely to vote Republican than are younger people
   d. your own:

2. Describe the specific behaviors or statements made by people that might lead to the observation you selected.

3. Based on the discussions and examples in the chapter, derive (make up) a reasonable theory for your observation. It should be more general than your observation.

EXERCISE 2.2

Name _____

Babbie describes the major social scientific paradigms. Identify and briefly summarize three major paradigms in one of the other disciplines you have encountered in college. Examples of disciplines include history, political science, economics, and psychology. Please note that the paradigms you describe will most likely be different from those discussed in the text; you need to identify the paradigms appropriate to the particular discipline you choose. Explain your choice of discipline and paradigms.

Discipline:

Paradigm # 1:

Paradigm # 2:

Paradigm # 3:

EXERCISE 2.3

Name _____

Imagine that you are a social scientist who wishes to examine in greater detail the implications of the paradigms discussed by Babbie. Specifically, you wish to apply the conflict and the structural functionalism paradigms in a study of how your school works. Give examples of how each paradigm may affect what you examine in your study.

Conflict paradigm:

Structural functionalism paradigm:

EXERCISE 2.4

Name _____

Use one of the Web search engines to locate additional information on one of the theoretical paradigms Babbie discusses (see www.socsciresearch.com for a guide to sociology sites). Or visit Michael Kearl's Sociological Tour of Cyberspace for some sites on theory (www.trinity.edu/mkearl/).

1.   Identify the paradigm you selected.

2.   Note two additional conclusions you can make about this paradigm that either amplify points raised in the text or raise new points about this paradigm.

EXERCISE 2.5

Name _____

Review Babbie's analysis of the conflict paradigm. Using deductive theory construction, we could hypothesize that more successful people will be most satisfied with the way things are because they benefit more from current social arrangements. Use SPSS or another data analysis program as indicated by your instructor to test this observation by running a crosstabulation of HAPPY by PRESTIG3 (see Appendix 1). Be sure to put PRESTIG3 on top (columns) and HAPPY on the side (rows), and request column percentages.

1.   Present your table.

1.   Analyze the percentage differences to determine if the hypothesis is correct.

ADDITIONAL INTERNET EXERCISES

1. Access the Intute: Social Sciences site at http://www.intute.ac.uk/socialsciences/http://sosig.esrc.bris.ac.uk and click on "Sociology." Then click "Sociological Theory." Select a category that resembles one of the paradigms in the text. Read and summarize an article from that category.

2. Access the Marx/Engels Internet Archive site at http://www.marxists.org/archive/marx/ and click on "Subject Index." Select and summarize the information presented on a topic of your choice.

3. Visit Michael Kearl's site on sociological theory at http://www.trinity.edu/mkearl/theory.html and select two sites he notes. Indicate how each site's content connects with this chapter.

Chapter 3

The Ethics and Politics of Social Research

OBJECTIVES

1. Discuss why ethical issues are frequently not apparent to the researcher.

2. Describe and illustrate the ethical issues involved in the following: voluntary participation, no harm to subjects, anonymity and confidentiality, the researcher's identity, and analysis and reporting.

3. Describe the role of Institutional Review Boards.

4. Identify which of the ethical principles were violated in the Humphreys tearoom study.

5. Identify which of the ethical principles were violated in the Milgram shock study.

6. Describe two ways in which ethical and political concerns differ.

7. Summarize the link between objectivity and ideology.

8. Compare the positions on the issue that social science can (or cannot) and should (or should not) be separated from politics.

9. Illustrate how political issues exist in some of the research on sexuality and for the Census.

10. Identify the three main purposes of Babbie's discussion on politics.

OUTLINE

1. Ethical issues in social research.
   a. Voluntary participation
   b. No harm to the participants
   c. Anonymity and confidentiality
   d. Deception
   e. Analysis and reporting
   f. Institutional Review Boards
   g. Professional codes of ethics

2. Two ethical controversies
   a. Trouble in the tearoom
   b. Observing human obedience

3. The politics of social research
   a. Objectivity and ideology
   b. The politics of sexual research
   c. Politics and the Census
   d. Politics with a little "p"
   e. Politics in perspective

SUMMARY

Practicing social research involves two important considerations: ethics and politics. Social scientists often disagree on ethical and political questions because many such issues are not always immediately apparent, and few present clear and absolute solutions.

Social scientists generally agree on five basic principles in handling ethical concerns. Voluntary participation is the first principle. This principle is important because social research often represents an intrusion into people's lives and often requires that people reveal personal information about themselves. Social research often requires that people reveal such information to strangers—the researchers. Although other professionals such as physicians and lawyers also require such information and are strangers to the person, the social researcher cannot make the claim that revealing information is required to serve the personal interests of the respondent. But social scientists do argue that involvement in research may ultimately help all humanity.

Some research subjects such as students and prisoners believe that participation might benefit them personally, and in such situations the decision to participate may not be purely voluntary. A different problem is presented by the scientific norm of generalizability, which requires that research findings should be based on representative samples. Voluntary participation compromises this principle.

Social scientists should never injure participants, regardless of whether or not participants volunteered for the study. This second principle pertains primarily to psychological harm, which may result from asking people to reveal deviant behavior, unpopular attitudes, demeaning personal characteristics, and the like. This principle also applies when participants are asked to deal with aspects of themselves that they do not normally consider, a practice that may cause some agony. Subjects may also be harmed by the reporting of data, such as when recognition is possible. Increasingly, federal and other funding agencies require an independent evaluation of the treatment of human subjects, which may help the researcher identify inappropriate research strategies.

The third principle involves protecting the anonymity and confidentiality of research subjects. This principle is related to the previous one of protecting the subject from harm. A respondent is anonymous when the researcher cannot connect a given response with a given respondent. In a confidential study, the researcher is able to identify a given person's responses but promises not to reveal this identity. Confidentiality can be enhanced by using identification numbers instead of names.

Fourth, the researcher's identity can also be an ethical problem. Often it is useful and even necessary to identify oneself as a researcher. Other times it is possible and important to conceal one's research identity. Doing so may enhance the completeness and accuracy of the data collected. But doing so also involves a serious ethical dilemma since deceiving people is generally unethical. Social scientists sometimes handle this dilemma by identifying themselves as researchers but failing to inform the respondent of the true nature of the study, which also poses an ethical dilemma. It is important to debrief subjects after the study to minimize the effects of deception.

Fifth, ethical concerns enter in the analysis and reporting of data. Ethical obligations to colleagues dictate accurate reporting of the shortcomings and negative findings in a study. Also, social scientists should not make accidental findings appear to be the result of careful hypothesizing and theorizing.

Finally, carelessness and sloppiness may be an ethical problem. Not all research must produce positive results, but it should be conducted in such a way that promotes that possibility.

Federal law affects ethical issues in social research. Any agency which receives federal research support must create an Institutional Review Board. Such boards include faculty (and sometimes other professionals) who review research proposals involving human subjects to insure that the subjects' rights and interests are being protected. Some categories of research are exempt from IRB review, such as the use of existing data sets. Research of and through the internet has drawn ethical scrutiny lately, although the basic ethical principles already noted apply here as well. In addition, most social science professional associations have created and published formal codes of ethics.

Two studies have gained notoriety for the ethical issues they have raised: Laud Humphreys's participant observation study of homosexuality in public restrooms and Stanley Milgram's laboratory studies of obedience to authority. Humphreys's work has been attacked by people both within and outside the social scientific community for his invasion of the privacy of the homosexuals and for his deceit in obtaining their identities. Milgram's shock experiments have been criticized for the psychological suffering experienced by the participants.

Ethical issues are distinguished from political issues in two respects. The ethics of social research apply more to the methods employed, and political issues are more concerned with the substance and use of research. Second, there are no formal codes of acceptable political conduct comparable to the codes of ethical conduct that many professional associations have established.

One exception to the absence of political norms is the generally accepted stance that a researcher's personal political orientation should not influence the research completed or its interpretations. This agreement enhances objectivity by means of intersubjectivity, whereby social scientists with different subjective views arrive at the same results when they apply the same techniques. However, this notion of "value-free sociology" has come under attack in recent years as Marxist and neo-Marxist researchers have argued that social science should be clearly linked to social action. They argue that explanations of the current state of affairs inevitably results in defending the status quo.

There are numerous examples of social research intertwined with politics. Social scientific research has influenced American racial policies and has generated political and ideological debates within the social science community. Research on sexuality has generated particular political scrutiny. Social science research is closely connected with political concerns. Even the seemingly simply matter of counting people for the Census is politically charged. The "politicization of science" is a recent hot-button issue as some feel that science is a threat to religious views and others feel that concentrated political power may threaten the independent functioning of scientific research.

It is also clear that social science research proceeds even under political controversy and hostility. It is important to make ideological considerations a part of the backdrops we create, which will improve our awareness as we study research methods. Researchers should participate in public debates and express both their scientific expertise and their personal values.

TERMS

1. anonymity
2. code of ethics
3. confidentiality
4. debriefing
5. deception
6. informed consent
7. Institutional Review Boards
8. value-free sociology
9. voluntary participation

MATCHING

_____ 1. An ethical principle that presupposes the willingness of respondents to participate in social research.

_____ 2. The condition that exists when the researcher cannot identify a given response with a given respondent.

_____ 3. The condition that exists when the researcher is able to identify a given person's responses but promises not to do so publicly.

_____ 4. Published formal guidelines for conducting social research.

_____ 5. Sociology that is relatively unaffected by personal values.

_____ 6. A mechanism for reviewing ethical issues within an institution.

_____ 7. Interviews conducted following an experiment to discover any problems generated by the research experience so that those problems can be addressed.

_____ 8. Not being truthful about your identity as a researcher.

_____ 9. The formalized process for securing voluntary participation.

TRUE-FALSE QUESTIONS

T F   1. The ethical principle that subjects should be free to participate in an study or not is known as the no harm to participants principle.

T F   2. Informed consent pertains to the ethical norms of voluntary participation and no harm to participants.

T F   3. With confidentiality, a researcher has no way of knowing the identity of research participants.

T F   4. Debriefing is commonly used to counteract the effects of deception.

T F   5. In addition to their ethical obligations to subjects, researchers have ethical obligations to their colleagues in the scientific community as well.

T F   6. The acceptable research behaviors of researchers as published by professional associations are known as rules of engagement.

T F   7. Ethical considerations are almost always apparent to us.

T F   8. Negative findings are less important than positive findings and should not be reported.

T F   9. The fact that people in social research often reveal personal information about themselves to researchers, who are strangers, is most closely related to the ethical principle of voluntary participation.

T F   10. When Milgram asked research subjects to administer "shocks" to "pupils" who were confederates of the researcher, the subjects generally refused to continue administering shocks.

REVIEW QUESTIONS

1. Ethical considerations enter at which point in the research process?
   a. selection of topic
   b. selection of subjects
   c. data gathering
   d. data analysis
   e. at all stages

2. The topic of ethics is typically associated with
   a. humanity.
   b. religion.
   c. medicine.
   d. psychology.
   e. morality.

3. The *major* justification the social scientist has for requesting participation in a study is that
   a. it may help the respondent.
   b. it may help all humanity.
   c. it may help the social scientist.
   d. it may help government officials make policy decisions.
   e. it may help improve the educational system.

4. The ethical principle of voluntary participation *most* threatens which scientific goal?
    a. parsimony
    b. objectivity
    c. cumulativeness
    d. generalizability
    e. intersubjectivity

5. For her senior project in political science, Cheri conducts a survey on students' attitudes and behavior concerning homosexuality. While distributing the questionnaire, she assures the group of students that no one will be able to trace responses to an individual. However, she has obtained a seating chart with the names of all the students in the class and where they were sitting. Cheri is violating which ethical principle?
    a. confidentiality
    b. anonymity
    c. harm to subjects
    d. concealed identity of researcher
    e. voluntary participation

6. The controversy surrounding Milgram's study involving "shocks" suggests he *most* violated which ethical principles?
    a. harm to subjects and concealed purpose of study
    b. anonymity and harm to subjects
    c. confidentiality and harm to subjects
    d. concealed purpose of study and anonymity
    e. concealed purpose of study and confidentiality

7. The controversy surrounding Laud Humphreys's study of homosexuals suggests he *most* violated which of the following ethical principles?
    a. anonymity and confidentiality
    b. harm to subjects and data reporting without identification
    c. concealed identity of researcher and anonymity
    d. value-free inquiry and concealed identity of researcher
    e. harm to subjects and anonymity

8. What is considered ethical is based on
    a. a list of moral absolutes.
    b. the authority of the state.
    c. agreed upon principles within a community.
    d. cultural directives.
    e. codes of political behavior.

9. Federally mandated Institutional Review Boards were designed *primarily* to address which one of the following ethical principles?
    a. no harm to participants
    b. confidentiality
    c. anonymity
    d. voluntary participation
    e. identification of research purpose

10. Which of the following is *not* a difference between ethical and political aspects of social research? Or are they all differences?
    a. Ethical considerations are more objective than political considerations.
    b. Ethical aspects include a professional code of ethics, whereas political aspects do not.
    c. Ethics deal more with methods, whereas political issues deal with substance.
    d. Ethical norms have been established, whereas political norms, for the most part, have not been established.
    e. All are differences.

11. Recently, Marxist scholars have argued that social research should be closely connected to social action. Which feature of science is *most* threatened by this position?
    a. intersubjectivity
    b. objectivity
    c. parsimony
    d. cumulativeness
    e. generalizability

12. The politics of research centers on the issues of
    a. voluntary participation and anonymity.
    b. anonymity and no harm to subjects.
    c. no harm to subjects and objectivity.
    d. objectivity and ideology.
    e. ideology and voluntary participation.

13. Josephine is a student at Consolidated Community College and is planning to do a study on how eating disorders affect academic performance. Who must review her research before it can be done?
    a. the federal government
    b. the state government
    c. the American Sociological Association
    d. the Institutional Review Board
    e. her department chair

14. The discussion of the Census best illustrates which of the following points?
    a. the government needs to get more people to respond to the Census surveys
    b. undercounting segments of the population has political implications
    c. the government needs to better address the ethical issues, particularly voluntary participation
    d. Census workers need to be better trained
    e. anonymity is not preserved very well in Census surveys

15. The only partial exception to the lack of political norms is the generally accepted view that
    a. governments should not dictate codes of ethics.
    b. a researcher's personal political orientation should not interfere with or influence the research.
    c. politics should not be discussed in social research because no one can agree on what should be done.
    d. professional associations should incorporate more about politics in their codes of ethics.
    e. researchers should get training in how politics affects Institutional Review Boards

16. Bernetta did a study in which she, the researcher, could not identify a given response with a given respondent. She employed
    a. anonymity.
    b. deception.
    c. value-free sociology.
    d. confidentiality.
    e. ideological bias.

17. Debbie did an experiment and was particularly concerned that her subjects base their voluntary participation on a full understanding of the possible risks involved. This is an example of
    a. a code of ethics.
    b. an Institutional Review Board.
    c. informed consent.
    d. deception.
    e. debriefing.

18. Cameroon did a study in which he could identify a given person's responses but promised not to do so publicly. He employed
    a. anonymity.
    b. deception.
    c. value-free sociology.
    d. confidentiality.
    e. ideological bias.

19. Lying is particularly common in which one of the following research designs?
    a. surveys
    b. laboratory experiments
    c. field experiments
    d. field research
    e. nonparticipant observation

20. Regarding the reporting of study shortcomings, researchers should
    a. tell readers about the shortcomings.
    b. include them only in oral presentations.
    c. make it look like there were not any shortcomings in the study.
    d. explain why the shortcomings were really no big deal.
    e. bury them in a footnote.

DISCUSSION QUESTIONS

1. Summarize the ethical agreements discussed in the text. Which one do you think is the most important? Why?

2. Discuss which of the ethical agreements were violated in the Humphreys study of homosexuality and in the Milgram study of obedience.

3. Discuss the similarities and differences between the political and ethical aspects of social science research.

EXERCISE 3.1

Name _____

Listed below are research situations that involve ethics. You are to rank order these situations in terms of how seriously they violate the ethical agreements discussed in the chapter and to explain the reasons for your rankings.

a. A psychology instructor asks students in an introductory class to complete questionnaires that the instructor will analyze and use in preparing a journal article for publication.

b. After a field study of deviant behavior during a riot, law enforcement officials demand that the researcher identify those persons who were observed looting. Rather than risk arrest as an accomplice after the fact, the researcher complies.

c. After completing the final draft of a book reporting a research project, the author discovers that 25 of the 2,000 survey interviews were falsified by interviewers but chooses to ignore that fact and publish the book anyway.

d. A Ph.D. candidate gains access to an underground mine as a researcher-employee by telling management that he has a BA degree and wants to get some practical experience before going on for a degree in metallic engineering. He is very familiar with mining work.

e. A college instructor wants to test the effect of unfair berating on exam performance. She administers an exam to both sections of a course. The overall performance of the two sections is essentially the same. The grades of one section are artificially lowered and the instructor berates the students for performing so badly. She then administers the same final exam to both sections and discovers that the performance of the unfairly berated section is worse. The hypothesis is confirmed, and the results are published.

f. In a study of sexual behavior, the investigator wants to overcome subjects' reluctance to report what they might regard as deviant behavior. To get past their reluctance, subjects are asked, "Everyone masturbates now and then; about how often do you masturbate?"

(continued)

g. A researcher discovers that 85 percent of students smoke marijuana regularly at a school. Publication of this finding will probably create a furor in the community. Because no extensive analysis of drug use is planned, the researcher decides to ignore the finding and keep it quiet.

h. To test the extent to which people may try to save face by expressing attitudes on matters they are wholly uninformed about, the researcher asks for their attitudes regarding a fictitious issue.

i. A research questionnaire is circulated among students as part of their university registration packet. Although students are not told they must complete the questionnaire, the hope is that they will believe they must, thereby ensuring a higher completion rate.

j. A researcher promises participants in a study a summary of the results. Later, due to a budget cut, the summary is not sent out.

k. A professor rewrites a thesis of one of his or her graduate students into an article and lists himself/herself as the sole author.

l. A panel of reputable social scientists is pressing for Congressional approval of a National Data Service, which would combine all data on particular individuals from separate data files into one master data file. The advantages of such a center would be reduced duplicity of efforts and an increased number of variables per individual.

m. A graduate student joins an internet discussion board about the sports car that she drives. She occasionally posts some responses on the board and decides to do a sociological study about the sense of community experienced by members. She attends a car rally sponsored by the board and mingles with participants. She uses archived discussion posts and notes she kept while at car rallies for her study.

(continued)

1. For each of the above issues, identify what you believe to be the one or two ethical principles that are most apparent in the situation. Explain why.

a:

b:

c:

d:

e:

f:

g:

h:

i:

j:

k:

l:

m:

(continued)

2.   Place each situation in one of the following three groups:

Minor ethical violations	Moderate ethical violations	Severe ethical violations

3. Explain the criteria that you used in your rankings.

EXERCISE 3.2

Name _____

Listed below are some goals and techniques associated with "ethical" social science research. Describe briefly how each might conflict with one or more of the "scientific" norms of social research discussed in the first few chapters. Use real or hypothetical research examples to illustrate your answers. For each ethical norm be sure to describe a specific scientific norm that it would conflict with, and be sure to present a real or hypothetical example of how each conflict might occur and what the conflict would be about. Also, for each conflict, indicate which norm—the ethical or the scientific—you would decide to go along with and explain why you made the decision you did.

1. Voluntary participation.
   a. How it conflicts with a scientific norm:

   b. Example of conflict:

   c. Which norm you would follow and why:

2. Anonymity and confidentiality.
   a. How it conflicts with a scientific norm:

   b. Example of conflict:

   c. Which norm you would follow and why:

(continued)

3. No harm to subjects.
   a. How it conflicts with a scientific norm:

   b. Example of conflict:

   c. Which norm you would follow and why:

4. Identifying yourself as a researcher.
   a. How it conflicts with a scientific norm:

   b. Example of conflict:

   c. Which norm you would follow and why:

5. Identifying the sponsor of your research.
   a. How it conflicts with a scientific norm:

   b. Example of conflict:

   c. Which norm you would follow and why:

EXERCISE 3.3 Name _____

Visit the Web site of the American Sociological Association (www.asanet.org, click "Ethics" on the left, which will take you to the ASA's code of ethics. Or visit the site for the British Sociological Association (www.britsoc.co.uk/ and click "Equality and Ethics" on the bottom right, click "Statement of Ethical Practice" under "Resources") and review the code of ethics.

1. Cite the language that addresses each of the ethical issues Babbie addresses: voluntary participation, no harm to subjects, anonymity and confidentiality, the researcher's identity, and analysis and reporting.

2. Summarize the code's content about the politics of social research.

3. Identify three other topics addressed in the code and summarize the content.

EXERCISE 3.4

Name _____

Review the General Social Survey in Appendix 1 of this volume, including the introduction. Determine how adequately this study addresses each of the ethical issues Babbie addresses: voluntary participation, no harm to subjects, anonymity and confidentiality, the researcher's identity, and analysis and reporting.

1.   Voluntary participation.

2.   No harm to subjects.

3.   Anonymity and confidentiality.

4.   Researcher identity.

5.   Analysis and reporting.

ADDITIONAL INTERNET EXERCISES

1. Click on "research ethics" at http://www.ethicsweb.ca/resources/ and click "Research Ethics." Read the Belmont Report and one other article/report that interests you. Briefly summarize both readings and describe how these readings support or contradict Babbie's discussion of ethics in research.

2. Review the Institute for Global Ethics website at http://www.globalethics.org/ and discuss the ethical issues raised in the Ethics Newsline Commentary. Describe what other resources are available on the site.

3. Visit www.sciencedaily.com/releases/2007/09/070911073925.htm to read about ethical and political issues encountered in research in the developing parts of the world. Show three connections with this chapter.

Part 2

The Structuring of Inquiry

Chapter 4

Research Design

OBJECTIVES

1. Identify the two major aspects of research design.

2. Define and illustrate the three basic purposes of research.

3. List three reasons for performing exploratory studies.

4. Describe the three main criteria for nomothetic, causal relationships.

5. Describe the three false criteria for nomothetic causation.

6. Differentiate necessary and sufficient causes.

7. Define units of analysis and identify and illustrate each of the basic types.

8. Define and illustrate the ecological fallacy.

9. Define and illustrate reductionism.

10. Select four possible research issues, and identify the unit of analysis for each.

11. Compare cross-sectional and longitudinal studies in terms of the advantages and weaknesses of each.

12. Differentiate among the three types of longitudinal studies–trend, cohort, and panel–by definition and example.

13. Explain how longitudinal studies may be approximated using the cross-sectional design

14. Depict the research process in a diagram and describe the diagram.

15. Identify the basic elements of a research proposal.

16. Identify the ethical issues in research design.

OUTLINE

1. Three purposes of research
   a. Exploration
   b. Description
   c. Explanation

2. The logic of nomothetic explanation
   a. Criteria for nomothetic causality
   b. False criteria for nomothetic causality

3. Necessary and sufficient causes

4. Units of analysis
   a. Individuals
   b. Groups
   c. Organizations
   d. Social interactions

SUMMARY

Research design involves developing strategies for executing scientific inquiry. It involves specifying precisely what you want to find out and determining the most efficient and effective strategies for doing so. Appropriate research designs enable the social scientist to make observations and interpret the results.

Social scientists typically have one or more of the following as goals for their research: exploration, description, and explanation. Exploratory studies are often done when a researcher is examining a new interest, or when the subject of study is relatively uncharted. They help to determine the feasibility of a larger scale study and to develop the methods for such a study. The researcher's intent in a descriptive study is to observe and describe some segment of social reality. Explanatory studies are undertaken to identify possible causal variables of a given social phenomenon, thereby contributing to understanding.

Recall that nomothetic explanations identify a few factors (independent variables) that can explain much of the variation in a particular dependent variable. There are three main criteria for establishing nomothetic, causal relationships in social science research. First, there must be a correlation between the two variables. That is, as the independent variable changes, the dependent variable changes. Second, the time order must be clear: the independent variable must come before the dependent variable. Third, the relationship must be nonspurious. That is, the observed relationship between the independent and dependent variables cannot result from the impact of a third variable. Social scientists use the nomothetic model of causal analysis in hypothesis-testing by specifying the strength of the relationship expected and by specifying tests for spuriousness that any observed relationship must survive.

Three caveats apply to these criteria by way of false criteria. The first is complete causation: social scientists speak in terms of probabilistic causation, not complete causation. That is, a pattern exists, but that pattern does not apply to everyone studied. The second is exceptional cases: exceptions to the pattern do not disprove a causal relationship. The third is majority of cases: we focus on relative differences by comparing the categories of the variables to see if there is a difference. Even though all the categories may show relatively low or high scores, one category may still be somewhat different from the rest.

Knowing about necessary and sufficient causes may help us better understand false criteria for causality. A necessary cause represents a condition that must be present for the effect to follow. A sufficient cause represents a condition that, if it is present, guarantees the effect in question. Note that a sufficient cause is not the only possible cause of a particular effect. Note also that a cause can be sufficient, but not necessary. In the social sciences we never discover single causes that are absolutely necessary and absolutely sufficient when analyzing the nomothetic relationships among variables. However, we do occasionally find causal factors that are either 100-percent necessary or 100-percent sufficient. Idiographic causes are sufficient but not necessary.

Social scientists typically study individual people as their units of analysis, although they do so in aggregate form. But they also frequently study social groups, such as families. Even though the unit of analysis is the group, characteristics of the group may be derived from the characteristics of individual members. Formal social organizations, like a corporation, are also a unit of analysis. Sometimes social interactions are the unit of analysis, where the focus is what goes on among humans. Finally, social artifacts are the products of social beings or their behavior and can also be analyzed. Realize that many other types of units of analysis are possible.

It is critically important to identify the units of analysis in a study accurately. Failure to do so may result in committing the ecological fallacy: gathering data on one unit of analysis (typically groups) but making assertions regarding another (typically individuals). While you strive to avoid committing this error, remember that generalizations and probabilistic statements are not invalidated by individual exceptions. Reductionism is another type of faulty reasoning connected to units of analysis. Reductionism involves attempts to explain a particular phenomenon in terms of limited and/or lower-order concepts. Sociobiology is an example of reductionism.

Another aspect of research design is the time dimension. The major distinction lies between cross-sectional studies (observations at one point in time) and longitudinal studies (observations made at multiple time points). There are three types of longitudinal studies. Trend studies are those that examine changes within a population over time. Cohort studies examine more specific subpopulations as they change over time. Panel studies examine the same set of people over time. Longitudinal studies are clearly superior for making causal assertions, but approximations to this design are frequently found in cross-sectional studies through such techniques as using simple logic, establishing clarity of time order of the variables, asking retrospective questions, and making age-cohort comparisons.

The research process begins with an interest, idea, or theory, which may be more exploratory in nature. Clearly define the purpose of your study. Pursue clear conceptualization and operationalization. Appropriate research designs and sampling strategies must be used. The product is a series of observations (data gathering), which are processed and analyzed, and the results are linked with the original framework of the study. Potential applications of the results should be considered. Although the research process does not always proceed in such a neat fashion, all these elements must be addressed. In designing a research project, researchers find it useful to assess their own interests, their abilities, and the resources available. Some find that multiple research strategies are the most effective, a process known as triangulation.

Research proposals are used to describe a potential study explicitly. They involve a clear statement of the problem, a literature review, a description of the subjects, a description of the measurement and data collection methods, an outline of the intended analyses, a schedule, and a budget. Depending on the nature of your research design, you may need to submit it to your institution's Institutional Review Board to help protect human subjects. In any case, be sure to follow the basic ethical principles outlined in Chapter 3.

TERMS

1. cohort studies
2. complete causation
3. correlation
4. cross-sectional studies
5. description
6. ecological fallacy
7. exceptional cases
8. explanation
9. exploration
10. focus groups
11. idiographic explanation
12. longitudinal studies
13. majority of cases
14. necessary cause
15. nomothetic explanation
16. panel studies
17. reductionism
18. social artifacts
19. social interactions
20. sociobiology
21. spurious relationship
22. statistical significance
23. sufficient cause
24. trend studies
25. triangulation
26. units of analysis

MATCHING

_____ 1. Explaining a particular phenomenon in terms of limited and/or lower-order concepts.

_____ 2. The purpose of research that stresses the determination of causes.

_____ 3. The "what" or "whom" that is actually studied.

_____ 4. The error in which data are gathered from one unit of analysis but conclusions are made about another unit of analysis.

_____ 5. The purpose of research that stresses examining a new area of interest or studying uncharted areas.

_____ 6. Studies in which the units of analysis are studied at one point in time.

_____ 7. Longitudinal studies in which independent samples of the same age group are studied at two points in time or more.

_____ 8. Longitudinal studies in which the same set of individuals is studied at two or more points in time.

_____ 9. A paradigm based on the view that social behavior can be explained in terms of genetic characteristics and behavior.

_____ 10. The use of several different research methods to test the same finding.

TRUE-FALSE QUESTIONS

T   F   1.   A researcher who wishes to become familiar with a new area of interest should use an explanatory approach.

T   F   2.   Nonspuriousness is the requirement for a causal relationship that says that the effect cannot be explained in terms of some third variable.

T   F   3.   A necessary cause represents a condition that, if it is present, guarantees the effect in question.

T   F   4.   The most typical unit of analysis is the individual.

T   F   5.   The use of yearbooks as the unit of analysis is an example of social artifacts.

T   F   6.   Using data on cities to develop conclusions about individuals reflects reductionism.

T   F   7.   A review of how support for abortion has changed over a decade is an example of a cross-sectional study.

T   F   8.   It is possible to draw approximate conclusions about processes that occur over time by using cross-sectional data.

T   F   9.   Units of analysis are limited to individuals, groups, organizations, and artifacts.

T   F   10.   Exceptions frequently disprove a causal relationship in nomothetic explanations.

REVIEW QUESTIONS

1. Exploratory studies are done most often for purposes of
   a. predicting future behavior, testing feasibility of a more careful study, developing methods to be used in a more careful study.
   b. connecting empirical reality with theory, satisfying a researcher'sociology curiosity, developing methods to be used in a more careful study.
   c. satisfying a researcher'sociology curiosity, testing feasibility of a more careful study, developing methods to be used in a more careful study.
   d. connecting empirical reality with theory, predicting future behavior, developing methods to be used in a more careful study.
   e. satisfying a researcher'sociology curiosity, testing feasibility of a more careful study, connecting empirical reality with theory.

2. The degree of relationship between two variables is of particular concern in which type of study?
   a. exploratory
   b. descriptive
   c. explanatory
   d. qualitative
   e. panel

3. Researcher Lewis spends three months as a choir member touring Europe to see what off-stage choir member interaction is all about. This is an example of which kind of study?
   a. cohort
   b. cross-sectional
   c. exploratory
   d. descriptive
   e. trend

4. Professor Johannison has just discovered that amount of time spent in social activities has a causal influence on academic performance. This is an example of which kind of study?
   a. explanatory
   b. exploratory
   c. cross-sectional
   d. descriptive
   e. trend

5. If a researcher wanted to find out how much time elderly persons in nursing homes spend interacting with each other, the researcher would conduct which kind of study?
  a. exploratory
  b. cohort
  c. explanatory
  d. descriptive
  e. time-lag

6. Which of the following *best* reflects the ecological fallacy?
  a. data gathered from individuals, conclusions drawn about individuals
  b. data gathered from individuals, conclusions drawn about groups
  c. data gathered from groups, conclusions drawn about organizations
  d. data gathered from groups, conclusions drawn about individuals
  e. data gathered from organizations, conclusions drawn about nations

7. Read the following statement and then determine the appropriate unit of analysis: "most Americans believe in God." What unit of analysis is reflected?
  a. individual
  b. group
  c. organization
  d. social artifact
  e. social interaction

8. Read the following statement and then determine the appropriate unit of analysis: "ten percent of families in a community move within a year." What unit of analysis is reflected?
  a. individual
  b. group
  c. organization
  d. social artifact
  e. social interaction

9. Read the following statement and then determine the appropriate unit of analysis: "church records show that contributions have declined over the past 20 years." What unit of analysis is reflected?
   a. individual
   b. group
   c. organization
   d. social artifact
   e. social interaction

10. A particular weakness in cross-sectional studies is that
    a. too often too few people are studied.
    b. too often too few variables are studied.
    c. they are limited to individuals as the units of analysis.
    d. they are too descriptive and insufficiently explanatory.
    e. they make it difficult to infer causality.

11. Professor Lodwicky used the 2000 United States Census and the 1990 Census to compare the average family size. This reflects which type of study?
    a. cross-sectional exploratory
    b. cross-sectional descriptive
    c. trend exploratory
    d. trend descriptive
    e. panel exploratory

12. Which type of study would *best* enable a researcher to assess changes associated with attending college for four years?
    a. cross-sectional
    b. longitudinal
    c. trend
    d. cohort
    e. panel

13. Which one of the following is ***not*** an appropriate strategy for approximating longitudinal studies when only cross-sectional data are available?
    a. using logical inference
    b. examining the time order of the variables
    c. examining differences in responses of those who return their questionnaires early versus late
    d. examining age differences
    e. asking respondents to recall their pasts

14. The research proposal generally begins with
    a. a literature review.
    b. a statement of a problem or objective.
    c. identifying subjects for study.
    d. measurement.
    e. data collection.

15. Which of the following ***best*** reflects the nomothetic model of explanation?
    a. Professor Guys wishes to explain alienation with causal variables.
    b. Professor Owings wishes to identify all or almost all the reasons why people retire early.
    c. Professor Rumpley wishes to perform an exploratory study of doctor-nurse interactions.
    d. Professor Wong wishes to identify the major causes of religiosity.
    e. Professor Hmong wishes to extend the generalizability of his grounded theory analysis of death anxiety.

16. Which one of the following hypotheses ***best*** fits the three criteria of causality?
    a. An increase in status attainment causes an increase in age.
    b. An increase in educational attainment causes an increase in status attainment in occupation, controlling for several variables.
    c. An increase in drowning causes an increase in ice cream consumption.
    d. An increase in conservatism causes a religious conversion, controlling for several variables.
    e. More males than females frequent bars.

17. Senator Robertson wants to know how voter interest in issues in his district has changed over time. Which design would be the *best* to use?
    a. trend
    b. cohort
    c. panel
    d. cross-sectional
    e. experimental

18. Nate wants to make some causal assertions in his cross-sectional study about the effects of high school involvement on violence among young adults (all the people he studied are in their early 20s). He can approximate a longitudinal study by
    a. using path analysis.
    b. using age differences.
    c. following the same individuals over time.
    d. making logical inferences based on the time order of the variables.
    e. improving the sampling design.

19. Attrition of study participants is most critical for which of the following designs?
    a. cross-sectional
    b. trend
    c. cohort
    d. panel
    e. triangulation

20. Andrew studied family conflict as portrayed in the media by reading letters to advice columns. What was his unit of analysis?
    a. individuals
    b. groups
    c. organizations
    d. social interactions
    e. artifacts

DISCUSSION QUESTIONS

1. Discuss the reasons why a social scientist would do an exploratory study, a descriptive study, and an explanatory study. Select a research topic and show how this topic would be addressed in terms of each of the three purposes of research.

2. Professor Nelson is doing a study on the effects of the following variables on the drinking behavior of college students: class level, gender, number of high school extracurricular activities the student participated in, drug usage in high school, family stability during early adolescence, and religiosity. She has a grant to study 800 college students at one point in time and wishes to make some causal assertions. Discuss how she can approximate a longitudinal study, and discuss the types of causal assertions she can legitimately make. Also discuss the limitations on these assertions. Be specific, and use some or all of the variables in your answer.

3. Explain why causes that are both necessary and sufficient are difficult to identify in the social sciences.

4. A fellow student who has not taken research methods comes to you for assistance in developing a research project. Which of the research design issues discussed in this chapter would you suggest he address? Why? In what order? Why? What would be your response to his question: "Which of these is the most important?" Why did you respond as you did?

EXERCISE 4.1   Name _____

Make up examples for the following units of analysis. Do not use examples
from the text, the review questions, or this paragraph. Students often confuse
groups and organizations. Think of groups as smaller units than organizations,
although in both instances you are studying the collectivity instead of the
individuals within the collectivity. Hence a family would be a group and a
corporation would be an organization. Also, groups have a specific group
boundary; families are groups but women or blacks are categories of
individuals. Finally, people typically generally know all or most of the other
members in a group but not in an organization. So, a local sorority would be a
group but the national sorority would be an organization. The adult Sunday
School class would be a group but the larger church would be an organization.

1. Give three examples of topics that involve studying individuals as the unit
   of analysis.

2. Give three examples of topics that involve studying groups as the unit of
   analysis.

3. Give three examples of topics that involve studying organizations as the
   unit of analysis.

(continued)

4. Give three examples of topics that involve social interactions as the unit of analysis.

5. Give three examples of topics that involve studying social artifacts as the unit of analysis.

EXERCISE 4.2

Select a topic of interest. Select just one variable. This variable should be something that can vary over time and that is related to the variable age. Examples might include: voting in an election or level of prejudice. Think of other examples. Then show how that variable could be studied in each of the four research designs: cross-sectional, trend, cohort, and panel. Explain the designs for your variable.

Your variable:

Cross-sectional study:

Trend study:

Cohort study:

Panel study:

EXERCISE 4.3

Name _____

A key focus of this chapter is causality—identifying factors (independent variables) that affect other factors (dependent variables). Your assignment is to practice identifying these variables. Be sure to select sociological variables. Avoid examples used in this workbook or in the textbook, or those used by your instructor.

For example, my causal hypothesis might be, "the higher the incidence of poverty in a community, the higher the crime rate." I am assuming that poverty is a cause of crime. In order to answer question 5, I am going to have to show and describe how the three criteria for causation can be applied to my example. The alternative arguments or factors that should be examined before accepting your causal hypothesis refer to the third requirement for a causal relationship— the correlation cannot be explained away by some third variable. For example, perhaps I should examine education; low education can cause poverty, and also low education may prevent people from getting jobs and that may cause them to turn to crime. Hence, there may be no real relationship between poverty and crime once education is controlled. Or perhaps I should examine the proportion of minorities in a community. If discrimination against minorities exists, then the more minority people, the greater the poverty. Also, police may be more likely to arrest minorities. These variables may account for the relationship between poverty and crime in my hypothesis.

1. Identify a dependent variable—some event, behavior, or attitude that varies from time to time and whose variation you wish to explain. Make sure it is a sociological variable.

2. Identify an independent variable whose variation you believe might explain variation in the dependent variable you described above. Make sure it is a sociological variable.

(continued)

3. State a causal hypothesis for your independent and dependent variables.

4.   Explain why you expect the relationship noted in # 3.

5. Describe the three criteria that need to be met before you can reasonably argue that there is a causal relationship between your two variables. In your discussion, note at least one alternative argument or factor that should be examined before accepting your causal hypothesis (this reflects the third criterion). Apply this alternative factor to your specific hypothesis. Remember that an alternative variable must be related to both your independent and dependent variables such that it explains away the relationship between the independent and dependent variables, in the same way that the size of a fire explains away the relationship between the number of fire trucks on the scene and the amount of damage that occurs. Show how your alternative variable is related to both your independent and dependent variables.

EXERCISE 4.4

Name _____

Find a research article of interest in one of the social science journals in your library, or select one from the reader you are using in this or another social science course. Alternatively, review one or more of the many newspapers on the Web (www.newspaperlinks.com). Your instructor will tell you which of these options to use, or may give you an article or list of articles from which to select. Make a copy or bookmark your entry on your browser because this article may be used in later exercises; attach the copy to this exercise if requested to do so by your instructor.

1. Identify the unit(s) of analysis and explain your answer.

2. Identify the independent and dependent variables and explain your answer.

3. Identify which of the following designs were employed and explain your answer: cross-sectional, trend, cohort, or panel.

EXERCISE 4.5   Name _____

The General Social Survey contained in Appendix 1 contains data for 1994 and 2004. Hence, it allows you to perform a longitudinal analysis–a trend analysis. Use SPSS or another data analysis program as indicated by your instructor to examine how attitudes toward abortion have changed over those time periods. Run a crosstabulation of ABANY (approve abortion for any reason) by YEAR. Be sure to put YEAR on top (columns) and ABANY on the side (rows), and request column percentages.

1.   Present your table below.

2.   Summarize the percentage differences across the years.

3.   Develop an explanation for the changes.

EXERCISE 4.6

Name _____

Babbie discusses the three criteria for establishing nomothetic causality. Your assignment is to apply these criteria to a crosstabulation of attitudes regarding the legalization of marijuana by gender. Use SPSS or another data analysis program as indicated by your instructor to run a crosstabulation of attitudes regarding the legalization of marijuana (GRASS) by gender (SEX) (see Appendix 1). Be sure to put SEX on top (columns) and GRASS on the side (rows), and request column percentages.

1.   Present your table below.

2.   Summarize the relationship as reflected in the percentage differences.

1.   Apply the three criteria for establishing causality to the relationship between support for capital punishment and gender.

ADDITIONAL INTERNET EXERCISES

1. Visit http://sda.berkeley.edu/archive.htm and click the "GSS Cumulative Data File 1972-2004." Click "Frequencies or Crosstabs." Type ABANY in the row box and YEAR in the column box. Summarize what this trend analysis shows in terms of support for abortion. In this case, the item is support for abortion if the woman wants it for any reason.

2. Visit the Bureau of Labor Statistics site via http://stats.bls.gov/ and click on the section at the right entitled "Unemployment Rate" and then click "Historical Data." Examine monthly unemployment rates for the last ten years. What can you learn from examining one cross-sectional data point versus the longitudinal trend data?

3. Peruse the White House Social Science Briefing Room at http://www.whitehouse.gov/fsbr/ssbr. html and review two of the studies you find. Comment on the appropriateness of the research designs used in the studies.

Chapter 5

Conceptualization, Operationalization, and Measurement

OBJECTIVES

1. Restate the argument that anything that is real can be measured.

2. Define measurement and differentiate it from observation.

3. Link the terms conception, conceptualization, and concepts.

4. Differentiate among the following terms: direct observables, indirect observables, constructs, and concepts.

5. Illustrate reification and explain why it is an error.

6. Show how indicators and dimensions contribute to the conceptualization process.

7. Outline the logic behind the interchangeability of indicators.

8. Describe and compare real definitions, nominal definitions, and operational definitions.

9. Select three concepts and develop both nominal and operational definitions for each.

10. Show how the clarification of concepts is a key element in qualitative research.

11. Explain why definitions are more problematic for descriptive research than for explanatory research.

12. Explain why researchers must be clear about the range of variation in a concept that interests them.

13. Give examples of how the variation between extreme attributes of a variable can affect the operationalization of a concept.

14. Explain why attributes of a variable should be exhaustive and mutually exclusive, and give examples of each.

15. Differentiate the following four levels of measurement and give an example of each: nominal, ordinal, interval, and ratio.

16. Explain why it is important to know the level of measurement for the variables in a study.

17. Explain when single or multiple indicators should be used to reflect a concept.

18. Differentiate precision from accuracy by definition and example.

19. Define reliability and compare these strategies for improving the reliability of measures: text-retest method, split-half method, using established measures, and reliability of research workers.

20. Define validity and compare these types of validity: face validity, criterion-related validity, construct validity, and content validity.
    predictive

21. Describe the tension between reliability and validity.

22. Show how ethics enters into the measurement process.

OUTLINE

1. Measuring anything that exists
   a. Conceptions, concepts, and reality
   b. Conceptions as constructs

2. Conceptualization
   a. Indicators and dimensions
   b. The interchangeability of indicators
   c. Real, nominal, and operational definitions
   d. Creating conceptual order
   e. An example of conceptualization: the concept of anomie

3. Definitions in descriptive and explanatory studies

4. Operationalization choices
   a. Range of variation
   b. Variations between the extremes
   c. A note on dimensions
   d. Defining variables and attributes
   e. Levels of measurement
   f. Single or multiple indicators
   g. Some illustrations of operationalization choices
   h. Operationalization goes on and on

5. Criteria of measurement quality
   a. Precision and accuracy
   b. Reliability
   c. Validity
   d. Who decides what's valid?
   e. Tension between reliability and validity

6. The ethics of measurement

SUMMARY

Social scientists believe that anything that is real can be measured. This is accomplished through the twin processes of conceptualization and operationalization. Conceptualization involves specifying what is meant by a concept, while operationalization yields the specific measures. Social scientists also commonly distinguish measurement from observation. Observation is a more casual and passive activity while measurement involves careful and deliberate observations for purposes of describing objects and events in terms of the attributes of a variable. Most of the variables social scientists study do not actually exist in the same way that the book you are reading exists. Instead, they are made up and seldom have a single, unambiguous meaning.

Conception reflects the mental images we have of something. Coming to an agreement about what terms mean reflects conceptualization and results in a concept. Scientists measure three things. First, they measure direct observables–things we can observe simply and directly. Second, they measure indirect observables–less direct methods used to provide inferences about concepts. Third, they measure constructs–theoretical creations based on observations but that cannot be observed directly or indirectly. In short, concepts are constructs derived by mutual agreement from conceptions. We sometimes err by believing that the terms for constructs have intrinsic meaning, an error known as reification.

Conceptualization is the process through which we specify what we mean when we use particular terms. Concepts themselves are not real or measurable; the indicators of a concept provide this function. Indicators are signs of the presence or absence of whatever concept we are studying. Seldom can a single indicator capture the full richness of meaning in most concepts. As a result, social scientists develop multiple indicators arranged around specifiable aspects of a concept, known as dimensions. Multiple dimensions and multiple indicators contribute to a more sophisticated understanding of our concepts.

To the extent that multiple, interchangeable indicators produce similar findings, researchers can reach similar conclusions about general concepts, even if they disagree about definitions. This stance is based on the assumption that if different indicators all represent the same concept, then all of them will behave the same way that the concept would behave if it were real and observable.

Three types of definitions contribute to conceptual order. Real definitions reflect the essential nature of a concept but are not very useful in the measurement process. Conceptualization instead depends on nominal and operational definitions. A nominal definition is one that is assigned to a term in a given study, and an operational definition details the specific operations involved in measuring the concept. These elements of conceptualization underscore the process nature of this step and produce conceptual order so that communication and measurement can proceed. Interestingly, definitions are more problematic for descriptive research than for explanatory research because the goal of descriptive research is to describe something accurately.

Social scientists have a variety of options available regarding the measurement process. For one, they must decide on the range of variation appropriate for a given concept. This choice is influenced by the nature of the study and the expected distribution of attributes among the subjects under study. Another choice pertains to how fine the distinctions will be among the various possible attributes of a given variable. This choice is also influenced by the purpose of the study, and social scientists generally err on the side of gathering too much detail rather than too little when they are unsure about how to proceed.

Many times the selection of specific indicators for a variable is quite straightforward because such variables have obvious single indicators (such as age). Other variables are better operationalized through composite measures that combine a number of indicators reflecting separate dimensions (such as religiosity). In either case, attributes for a variable need to be both exhaustive (a complete list of attributes) and mutually exclusive (no overlapping attributes).

Different variables may represent different levels of measurement. Nominal measures employ numbers for classification purposes only; their attributes have only the characteristics of exhaustiveness and mutual exclusiveness. Variables whose attributes may be logically rank-ordered are ordinal measures; the numbers employed reflect relatively more or less of the variable being measured. The actual distance separating attributes in interval measures can be described as being equal, but interval measures have no absolute zero. Ratio measures do have an absolute zero value, enabling the calculation of ratios. It is important to identify the levels of measurement of the variables because analytical techniques vary according to the level of measurement of a variable.

Social scientists apply several criteria to enhance measurement quality. They strive to develop measures that are both precise and accurate. They pay particular attention to reliability (consistency) and validity (actually measuring the concept of interest). Problems with both appear repeatedly in social research. Reliability can be enhanced by asking people only things they are likely to know, by asking the same information more than once (test-retest method), by randomly assigning items to two sets and comparing the responses in the two sets (split-half method), and by using measures known for their reliability. Consistent application of the research design is also mandatory, what is known as research worker reliability.

Validity can be enhanced through establishing face validity, whereby measures are compared with common agreements about a concept and with a logical understanding of a concept. Criterion-related validity (or predictive validity) is based on an external criterion. Construct validity is based on the way a measure relates to other variables within a system of theoretical relationships. Finally, content validity refers to the extent to which a measure covers the range of meanings included in a concept. Social researchers need to consider both their colleagues and their subjects as sources of agreement on the measurement of concepts. Researchers also need to be sensitive to the trade-offs between validity and reliability in their quest for sound measurements. Note too that it is unethical to deliberately attempt to slant your results by using a biased definition of your concept.

TERMS

1. accuracy
2. concept
3. conception
4. conceptualization
5. constructs
6. construct validity
7. content validity
8. criterion-related validity
9. dimension
10. direct observables
11. exhaustive
12. face validity
13. indicators
14. indirect observables
15. interchangeabilty of indicators
16. interval measure
17. measurement
18. mutually exclusive
19. nominal definition
20. nominal measure
21. operational definition
22. ordinal measure
23. precision
24. ratio measure
25. real definition
26. reification
27. reliability
28. split-half method
29. test-retest method
30. validity

MATCHING

_____   1. Careful, deliberate observations of the real world in order to describe objects and events in terms of the attributes composing a variable.

_____   2. The process through which we specify what we will mean when we use particular terms.

_____   3. Real, observable things that give evidence of the presence or absence of the concept we are studying.

_____   4. The definition that is assigned a term.

_____   5. The definition that specifies the operations involved in measuring a concept.

_____   6. The consistency or repeatability of a measure.

_____   7. The extent to which a measure accurately reflects the concept we wish to measure.

_____   8. A method for establishing reliability that involves splitting a set of items into two halves and seeing how strongly the two correlate.

_____   9. Theoretical creations that are based on observations but that cannot be observed directly or indirectly.

_____   10. The degree to which a measure relates to some external criterion.

TRUE-FALSE QUESTIONS

T   F   1. Conceptualization is the process through which we specify what we mean when we use particular terms.

T   F   2. An indicator is a specifiable aspect of a concept.

T   F   3. Definitions are more problematic for descriptive research than
                for explanatory research.

T   F   4. Attributes are composed of variables.

T   F   5. Sex is an example of a nominal level variable.

T   F   6. It is possible for less precise descriptions to be more accurate.

T   F   7. The split-half method of reliability involves splitting the
                respondents into two halves.

T   F   8. Criterion-related validity involves comparing a measure to an
                internal criterion in the measure.

T   F   9. Precision and accuracy are basically the same.

T   F   10. A ratio level of measurement can be represented as a nominal
                level of measurement, but a nominal level of measurement
                cannot be represented as a ratio level of measurement.

REVIEW QUESTIONS

1. Which one of the following is *false*? Or are they all true?
    a. The concept ethnicity represents the term ethnicity.
    b. Ethnicity is both a concept and a variable.
    c. Ethnicity as a variable is composed of attributes.
    d. Ethnicity may have different definitions.
    e. All are true.

2. The number of times one swims a week as used to measure physical
    fitness is an example of a/an
    a. operational definition.
    b. indicator.
    c. concept.
    d. dimension.
    e. construct.

3. Which of the following is *least* likely to be a dimension of marital happiness?
   a. shared financial decision making
   b. sexual compatibility
   c. similar child rearing ideas
   d. couple communication
   e. length of courtship

4. Professor Lewis spends a few weeks working through what she means by feminism. She decides that it includes such areas as equal treatment before the law, equality in the job market, and belief in the equality of men and women. She then develops specific questions for each of these three areas, such that the sum of the items creates a feminism score. In this example, which of the following reflects conceptualization?
   a. the whole process
   b. the term *feminism*
   c. the three basic areas
   d. the calculation of a feminism score
   e. the questions

5. Which of the following *best* reflects interchangeability of indicators?
   a. Joe uses multiple indicators and then deletes two and adds three on the basis of a test for reliability.
   b. Cathy decides to use several indicators of one of her variables to improve validity.
   c. Lisa decides to use several indicators of both her variables to improve validity.
   d. James argues that it really doesn't matter which of the indicators of his dependent variable is used because whatever relationship exists between the dependent and independent variables will appear with any of the indicators.
   e. Susan argues that indicators can be exchanged across dimensions and that the reliability and validity as well as the basic relationship will not be altered.

6. Professor Foxdire has been working with the concept of stress for several years and has worked hard at developing a precise measure of stress. He then works harder at discovering the real meaning of stress and finally decides that stress is real and his measure genuinely reflects it. What error has he made?
   a. provincialism
   b. interchangeability
   c. objectification
   d. ecological fallacy
   e. reification

7. In operationalizing approval for abortion, Professor Oliver is particularly concerned that the measure will reflect those opposed to abortion as well as those with varying levels of approval. Which operationalization choice is he confronting?
   a. range of variation
   b. number of dimensions
   c. number of indicators
   d. variation between the extremes
   e. interchangeability of indicators

8. Professor Wilson decides to study academic aspiration and, in line with the general thinking on the concept, specifies that she will treat it as the highest degree desired. This is an example of a/an
   a. categorical definition.
   b. nominal definition.
   c. real definition.
   d. empirical definition.
   e. operational definition.

9. Operational definitions are more problematic for descriptive research than for explanatory research because
   a. interpretations of descriptive statements directly depend on operational definitions used.
   b. interpretations of explanatory statements directly depend on the causal connection between concept and operational definition.
   c. in explanatory research operationalization of concepts is irrelevant as long as causality is established.
   d. descriptive statements demand less specificity than do explanatory statements.
   e. explanatory research deals primarily with relationships between variables and not operational definitions.

10. Read the following item and then answer the question that follows: "What is your student status?"
    a. First-year
    b. Sophomore
    c. Junior
    d. Senior
    e. Transfer

11. Which of the following *most* accurately describes this question?
    a. It is exhaustive but not mutually exclusive.
    b. It is mutually exclusive but not exhaustive.
    c. It is both exhaustive and mutually exclusive.
    d. It is neither mutually exclusive nor exhaustive.
    e. It is both reliable and valid.

12. Grade point average reflects which level of measurement?
    a. nominal
    b. ordinal
    c. linear
    d. ratio
    e. interval

13. The importance of levels of measurement lies in
    a. distinguishing qualitative from quantitative research.
    b. clarifying the conceptualization process.
    c. determining which statistical techniques are appropriate.
    d. clarifying the operationalization process.
    e. constructing measures of the same concept at different levels of measurement.

14. Which one of the following is *false*?
    a. Precision is more important in descriptive than in explanatory research.
    b. Precision is less important than validity.
    c. Precision parallels validity.
    d. Operationalization enhances precision.
    e. A measure can be precise without being accurate.

15. Professor Garcia decides to employ 60 interviewers in his study of teacher militancy. Which one of the following most likely will be the *major* measurement concern?
    a. epistemology
    b. validity
    c. reliability
    d. conceptual adequacy
    e. ability to interchange the indicators

16. Personnel Director Putler develops a measure of job satisfaction. She shows it to a couple of other personnel directors to make sure her scale measures what it purports to measure. She is seeking
    a. reliability.
    b. precision.
    c. face validity.
    d. accuracy.
    e. indicator correlation.

17. Professor Peoples develops a new scale to measure love. When given repeatedly, it generally yields similar results. When correlated with a well-validated love scale, it shows a low correlation. Which one of the following statements is correct?
    a. The scale is reliable but not valid.
    b. The scale is valid but not reliable.
    c. The scale is both reliable and valid.
    d. The scale is neither reliable nor valid.
    e. The scale is generalizable but not precise.

18. Marketing Analyst Harris tests the validity of her new customer satisfaction scale by comparing its relationship to other variables found to be related to customer satisfaction scales in previous studies. For example, if previous studies showed that older customers were more satisfied, she examined how age related to her own measure of customer satisfaction. She is using which type of validity?
    a. face
    b. criterion-related
    c. content
    d. construct
    e. multidimensional

19. Ignatius wanted to be sure he covered the full range of meanings in his measurement of happiness among college students. Given this, he was particularly concerned with
    a. construct validity
    b. content validity
    c. face validity
    d. criterion validity
    e. multidimensional

20. Elmer worked hard to establish the meaning of the term "prejudice" for his study on students. He consulted the literature and colleagues who have studied prejudice in order to come to an agreement about what the term means. Elmer engaged in
    a. reification
    b. dimensionalization
    c. conceptualization
    d. validity
    e. reliability

21. Nora developed a definition of adjustment to college in such a way that this concept represented what researchers in the field have come to agree upon as the concept's meaning. Nora developed a/an
    a. nominal definition.
    b. operational definition.
    c. real definition.
    d. conceptualization process.
    e. valid measure.

DISCUSSION QUESTIONS

1. Present a case that we can measure anything that exists. Use examples.

2. Describe the process of conceptualization and operationalization.

3. Discuss why it is important to consider the levels of measurement of variables in doing research.

4. Differentiate reliability from validity. Which is more important? Why?

EXERCISE 5.1   Name _____

Select one of the following concepts: religiosity, feminism, patient anxiety, or marital happiness. Go through the measurement process by completing the steps below. In the first question, describe the conceptualization process that you would go through in measuring the concept you selected. Explain how (i.e., where you would get your ideas on what to include in your definition) you would develop a nominal definition for it, how you would develop an operational definition of it, and how you would actually measure it. In the second question, be sure your nominal definition of the concept makes sense. Your indicators (in the third question) should be measures of the various characteristics or qualities that you have stated in your nominal definition; that is, your indicators should correspond to your nominal definition.

1. Describe the conceptualization process you would employ to measure this concept. Be specific. Be sure your response reflects the conceptualization process.

2. Provide a nominal definition of the concept.

3. Describe indicators you would use in developing your operational definition.

EXERCISE 5.2

Name _____

Grade point average (GPA) is often assumed to measure the intelligence of a student relative to that of other students. Give two reasons why GPA may *not* be reliable and two reasons why GPA may *not* be valid as a measure of the intelligence of college students. Make sure your reasons regarding reliability address the consistency or repeatability of the elements comprising GPA and your reasons regarding validity address the extent to which GPA measures intelligence.

Problems of reliability:

Problems of validity:

EXERCISE 5.3   Name _____

Consult your library's online resources to locate a social science journal the
library has online. Select one of the concepts in the article and describe the
conceptualization and measurement processes employed. Provide specific
examples of dimensions and indicators and of nominal and operational
definitions. Assess the reliability and validity of the measure, either by
summarizing how the researcher addressed both or by your own analysis if the
researcher did not address both.

Article citation (if required):

Concept:

Conceptualization process used:

Operationalization of concept:

Dimensions of the concept (if any):

Assess reliability of the measure:

Assess validity of the measure:

EXERCISE 5.4

Name _____

The concept of attitude toward pornography might be measured with the following four items. Respondents could be asked to indicate if they thought that:

    a. sexual materials lead to breakdown of morals
    b. sexual materials lead people to commit rape
    c. sexual materials provide an outlet for bottled-up impulses
    d. sexual materials provide information about sex

1. Speculate about the conceptualization process that produced these items. That is, speculate about the thinking process a social scientist might have gone through to develop these four items to measure attitude toward pornography.

2. Explain how three of the four types of reliability techniques could be applied to the items.

3. Explain how three of the four types of validity techniques could be applied to the items.

EXERCISE 5.5

Name _____

Sue is 20 years old and Mary is 40 years old. Write a simple statement regarding Sue's and Mary's ages that illustrates each of the levels of measurement.

1. Nominal:

2. Ordinal:

3. Interval:

4. Ratio:

EXERCISE 5.6

Name _____

The split-half method for establishing reliability involves splitting your items into two halves and then seeing how well the scores on those two halves correlate. This exercise involves determining how reliably three items in the General Social Survey measure attitudes toward gender roles. As a modified split-half approach, use SPSS or another data analysis program as indicated by your instructor to see how the three indicators correlate (See Appendix 1). That is, run crosstabulations of FECHLD with FEFAM, FECHLD with FEPOL, and FEFAM with FEPOL. Because we are not suggesting that one is an independent variable and the other a dependent variable, you can put whichever ones you wish on the top (columns) and side (rows). But request both column and row percentages to facilitate your analyses.

1.   Report the table for FECHLD with FEFAM.

2.   Report the table for FECHLD with FEPOL.

(continued)

3.   Report the table for FEFAM with FEPOL.

4.   Analyze the degree to which these three items reliably measure attitudes towards gender roles.

ADDITIONAL INTERNET EXERCISES

1. First conceptualize and operationalize "race." Then read the Census Bureau's report on race at http://www.census.gov/population/www/socdemo/race/racefactcb.html and answer the following questions after reading the report: 1) What changes did the Bureau make to the conceptualization and operationalization of race in the 2000 census? 2) Why did the Bureau make these changes? 3) Could the difference in operationalization produce different conclusions about race? 4) Does the Bureau's conceptualization and operationalization of race coincide with yours? If so, how? If not, what is different?

2. Visit http://www.gmac.com/gmac/thegmat/gmatbasics/whyrelyongmatscores.htm the at the web site of the Graduate Management Admissions Councilhttp://www. What can you conclude about the reliability and validity of the GMAT?

3. Use your library's online resources to locate a social science article that includes measurement of a concept. Note the full citation and assess the adequacy of the conceptualization and measurement process in terms of Babbie's advice.

Chapter 6

Indexes, Scales, and Typologies

OBJECTIVES

1. Link the contents of this chapter with the previous chapter on conceptualization, operationalization, and measurement.

2. List three reasons why composite measures are frequently used in social science research.

3. Differentiate index from scale by definition and example.

4. List the two reasons why scales are generally superior to indexes.

5. Describe two misconceptions regarding scaling.

6. List the four steps involved in creating an index.

7. Define and illustrate face validity, unidimensionality, and variance as criteria for selecting items.

8. Describe the rationale and application for employing bivariate relationships among items in index construction.

9. Describe the rationale and application of employing multivariate relationships among items in index construction.

10. Describe how items can be scored in index construction.

11. Describe five strategies for handling missing data in index construction.

12. Compare the rationale and application of item analysis and external validation as strategies for validating an index.

13. Describe the logic and procedures of the Bogardus social distance scale.

14. Describe the logic and procedures of Thurstone scales.

15. Describe the logic and procedures of Likert scaling.

16. Describe the logic and procedures of the semantic differential.

17. Describe the logic and procedures of Guttman scaling.

18. Explain the coefficient of reproducibility.

19. Explain and illustrate how typologies are used in social science research.

OUTLINE

1. Indexes versus scales

2. Index construction
    a. Item selection
    b. Examination of empirical relationships
    c. Index scoring
    d. Handling missing data
    e. Index validation
    f. The status of women: an illustration of index construction

3. Scale construction
    a. Bogardus social distance scale
    b. Thurstone scales
    c. Likert scaling
    d. Semantic differential
    e. Guttman scaling

4. Typologies

SUMMARY

Composite measures of variables—indexes, scales, and typologies—are created by combining two or more separate empirical indicators into a single measure. Social scientists employ composite measures for several reasons: single indicators are seldom sufficient to measure a complex concept adequately; multiple items extend the range of scores available; and composite measures are efficient strategies for handling multiple items.

Both scales and indexes are typically ordinal composite measures of concepts. Indexes are constructed by accumulating scores assigned to individual attributes, and scales are constructed through the assignment of scores to patterns of attributes. Hence a scale takes advantage of any intensity structure that may exist among the attributes and conveys more information than does an index. Be aware that whether the combination of data items results in a scale depends on the sample used. Also, the use of specific scaling techniques does not guarantee the creation of a scale. Indexes are more popular in social research because scales are difficult to construct from a particular sample.

The first step in constructing an index is selecting items. The items must have face validity (logical validity) and be unidimensional (represent only one dimension). Each item should contain some variance, although the amount may be greater for some items than for others. The next step is to examine the relationships among the items to determine if the items are related to each other empirically; this can be done with percentage tables or correlation coefficients. Bivariate analysis of the items is used to identify those items that should be eliminated either because they are unrelated to the other items or because they are so closely related as to be redundant. Multivariate analysis of the items allows the researcher to determine whether each item adds something to the index independently of the other items.

Next comes scoring the items in an index. The researcher must decide on the desirable range of index scores and must attain both a range of measurement in the index as a whole and an adequate number of cases at each point in the index. The researcher must also decide whether to weight the items equally or differentially; equal weighting is typically employed. The scoring process must include provisions for handling missing data. Several strategies are available for doing so, such as excluding those cases with missing data from the construction of the index, treating missing data as one of the available responses, assigning the middle value, or using proportions based on what is observed.

The last step in index construction is to test for validity. In item analysis, the researcher examines the extent to which the composite index is related to the individual items. External validation is employed by testing the index against other indicators of the same concept. External validation sometimes presents a dilemma. If the index is found to be unrelated to the external validator, it may be hard to decide which is invalid—the index or the other indicator.

The Bogardus social distance scale is used to measure people's willingness to associate with members of different groups. The scale contains an intensity structure by reflecting varying degrees of closeness tolerated. Hence, one can predict which items were selected on the basis of a single score.

When a logical structure among the indicators is not clear, Thurstone scaling may be employed to develop a format for organizing items that have at least an empirical structure among them. Experts are asked to assign numerical scores to many indicators of the same variable; this score reflects how strongly the item reflects the concept. Items on which there is little agreement are eliminated as ambiguous, and several of the remaining items are selected to represent each scale score from the lowest value to the highest value. Scores are generally assigned on the basis of the items selected that have the highest value. Thurstone scaling is infrequently used today because of the time required to construct the scale.

Likert scaling was originally designed to determine the relative intensity of different items. Likert scaling is seldom used today. However, the item format devised by Likert is very popular. Typically, it is used to create simple indexes which consist of a series of questionnaire items with a uniformly scored set of ordinal response categories expressing varying degrees of agreement. Composite scores are calculated by adding the individual item scores. Like the Likert format, the semantic differential asks respondents to choose between two polar opposite positions. Researchers using this procedure must determine the appropriate dimensions for a concept and then find two opposite terms to represent the polar extremes along each dimension.

Guttman scaling is also based on the fact that some items may prove to be "harder" indicators of a concept than others. An important objective of Guttman scaling is to maximize the reproducibility of response patterns from a single scale score. The coefficient of reproducibility reflects the percentage of original responses that could be reproduced by knowing the scale scores used to summarize them. Coefficients above .90 are necessary to conclude that a scale exists.

Whereas indexes and scales provide measures of a single dimension, typologies are often employed to examine the intersection of two or more dimensions. Typologies are very useful analytical tools and can be easily used as independent variables, although the fact that they are not unidimensional makes it difficult to analyze them as dependent variables.

TERMS

1. bivariate relationship
2. Bogardus social distance scale
3. coefficient of reproducibility
4. composite measure
5. external validation
6. face validity
7. Guttman scaling
8. index
9. index validation
10. item analysis
11. Likert scale
12. mixed types
13. multivariate relationship
14. response patterns
15. scale
16. scale types
17. semantic differential
18. Thurstone scaling
19. typology
20. unidimensionality
21. variance

MATCHING

_____ 1. A type of composite measure that summarizes and rank-orders several specific observations and represents some more general dimension.

_____ 2. A type of composite measure composed of several items that have a logical or empirical structure among them.

_____ 3. The criterion for choosing indicators that emphasizes the importance of selecting items that logically reflect the concept being measured.

_____ 4. The element of a composite measure that indicates that the measure reflects only one dimension.

_____ 5. A measure of internal validity that involves examining the extent to which the composite measure is related to (or predicts responses to) the items in the index itself.

_____ 6. The generalizability of the results of a measure.

_____ 7. A type of index in which response categories reflect intensity of agreement and that assumes that each item has the same intensity as the others.

_____ 8. A measure summarizing the intersection of two or more variables.

_____ 9. A value calculated in Guttman scaling that shows the predictive accuracy of the scale.

_____ 10. A type of scale that asks respondents to choose between two opposite positions.

TRUE-FALSE QUESTIONS

T   F   1. An index is constructed by assigning scores to patterns of responses.

T   F   2. It is a good idea to minimize the variance in responses to items in an index.

T   F   3. Item analysis is a method of index validation.

T   F   4. The Bogardus social distance scale involves having independent judges rate each item.

T   F   5. The Likert approach is used more commonly in indices than in scales.

T   F   6. The semantic differential asks respondents to choose between two opposite positions.

T   F   7. The coefficient of reproducibility refers to how accurately one can predict original responses knowing only the scale scores.

T   F   8. Typologies are particularly useful for examining one variable in depth.

T   F   9. Generally, it is preferable to give items equal weight in an index as opposed to different weights.

T   F   10. You cannot assume that a given set of items comprise a scale simply because it has turned out that way in an earlier study.

REVIEW QUESTIONS

1. The *major* reason for using composite measures is that
    a. they enhance reliability.
    b. they enhance external validity.
    c. they yield validity coefficients.
    d. they yield more comprehensive measures with a wider range of scores.
    e. they enhance the conceptualization process, thereby strengthening the theory.

2. Which of the following is *not* a composite measure? Or are they all composite measures?
    a. Scales
    b. Indicators
    c. Indexes
    d. Typologies
    e. all are composite measures

3. The key difference between scales and indexes is that scales
    a. are generally longer.
    b. require more conceptualization.
    c. take into account the different strengths of the indicators.
    d. are more valid and reliable.
    e. assist in building typologies.

4. Professor Leet has developed several items to measure self-concept and checks with a few experts to make sure that his items logically represent self-concept. Which criterion of item selection is he using?
    a. Unidimensionality
    b. face validity
    c. internal strength
    d. reproducibility
    e. variance

5. In performing bivariate analyses of items, which is the ***most*** desirable outcome?
    a. Some relationships should be low, but most should be moderate.
    b. None should be negligibly correlated, but most should be moderately correlated.
    c. All should be modertately strongly related.
    d. All but a few should be very strongly correlated.
    e. All should be strongly correlated.

6. Which of the following should occur after examining the bivariate relationships among items but before combining items into an index?
    a. calculating a coefficient of reproducibility
    b. validating the index
    c. examining multivariate relationships among items
    d. assessing the reliability of the items
    e. examining response patterns across the answer categories per item

7. The two basic decisions to be made in index scoring are
    a. range of scores and weighting of items.
    b. weighting of items and number of items.
    c. number of items and scalability.
    d. scalability and validity.
    e. validity and range of scores.

8. Professor Bigelow examines the extent to which each item in her fatalism index correlates with the index. Which strategy for index validation is she using?
    a. split-half
    b. parallel forms
    c. item analysis
    d. external validation
    e. face validation

9. Professor Strike wishes to develop a composite measure of adjustment to retirement such that an ordinal scale can be calculated based on items with the same or similar ranges of intensity. Which of the following would be *best*?
   a. semantic differential
   b. Guttman scale
   c. Bogardus social distance scale
   d. Thurstone scale
   e. Likert scale

10. Professor Howitzer wishes to develop a composite measure of teaching effectiveness by using several dimensions, each with several possible answers between two extremes. The polar extremes are to be labeled. Which of the following would be *best*?
   a. semantic differential
   b. Guttman scale
   c. Bogardus social distance scale
   d. Thurstone scale
   e. Likert scale

11. Professor Delaney wishes to construct a composite measure of job satisfaction but is particularly concerned that the items selected reflect equal intervals. He does this by having experts rate each of the items and then he selects 11 items to represent the range of scores. Which of the following did he use?
   a. semantic differential
   b. Guttman scale
   c. Bogardus social distance scale
   d. Thurstone scale
   e. Likert scale

12. Professor Feyen wishes to develop a composite measure of abortion support such that she can accurately predict which items a respondent agrees with based on the score. Which of the following would be *best*?
   a. semantic differential
   b. Guttman scale
   c. Bogardus social distance scale
   d. Thurstone scale
   e. Likert scale

13. The key difference between typologies and composite measures is that typologies
    a. are more valid.
    b. are more reliable.
    c. reflect interval and ratio levels of measurement and hence allow more powerful statistics.
    d. contribute more to conceptual clarification.
    e. are built with more than one variable.

14. One particular weakness of typologies is that
    a. they have little validity.
    b. they have little reliability.
    c. it is difficult to use them as dependent variables.
    d. it is difficult to use them as independent variables.
    e. they are difficult to analyze empirically.

15. Which one of the following is *not* a suggested way to handle missing data in index construction?
    a. if there are relatively few cases with missing data, you may decide to exclude them from the construction of the index and the analysis
    b. it is always best to simply exclude any cases with missing data
    c. you may sometimes have grounds for treating missing data as one of the available responses
    d. a careful analysis of missing data may yield an interpretation of their meaning
    e. all of these are good strategies for handling missing data

16. Item analysis is used for:
    a. assessing index reliability
    b. index validation
    c. improving item wording
    d. determining whether to use a scale or an index
    e. determining possible ethical issues involved

17. Serrina wanted to analyze the intersection between two variables–school type (public or private) and school location (urban or rural)–in her study of student values. Which approach should she use?
    a. semantic differential
    b. typology
    c. coefficient of reproducibility
    d. scale types
    e. index of variation

18. External validation is to internal validation as prediction is to:
    a. validity
    b. reliability
    c. the connection between the composite index and individual items
    d. the best strategy for handling missing data
    e. precision

19. Both scales and indexes are which level of measurement?
    a. nominal
    b. ordinal
    c. interval
    d. ratio
    e. real

20. Freda wants to develop a measure of support for the legalization of marijuana such that she can assign scores to patterns of responses, such that some items reflect a relatively weak degree of the variable while others reflect something stronger. Which one of the following should she use?
    a. index
    b. scale
    c. reliability
    d. validity
    e. typology

DISCUSSION QUESTIONS

1. Give three reasons for the frequent use of indexes and scales in social science research.

2. What do indexes and scales have in common? How are they different?

3. Describe the four steps in constructing an index: (1) selecting possible items, (2) examining their empirical relationships, (3) combining some items into an index, and (4) validating the index. Provide specific advice for completing each step.

4. Compare and contrast the following scaling procedures: Likert, Thurstone, and Guttman. For each, note purpose, advantages, disadvantages, and procedures.

5. Describe the utility of typologies.

EXERCISE 6.1

Name _____

You are to develop a semantic differential scale for a concept of interest. Be sure to follow the advice presented in the chapter.

1. Identify your concept.

2. List at least five dimensions that adequately reflect your concept. Remember, dimensions are different aspects of a variable. For example, religiosity has a belief dimension, a ritual dimension, an experiential dimension, and so forth.

3. Set up the items to resemble the format outlined in the text, and include both the polar opposite terms and the ratings in between.

4. Describe how you would score the results.

EXERCISE 6.2
Name _____

Develop a typology in an area of interest. Explain why you selected the two or more dimensions in your typology. Present your typology in a format like that in Table 6-4 in the text. Explain the typology and show how it could be used in a research study. Be specific.

As you construct your typology, be sure that each of the cells makes sense. For each cell in the table, be sure you can imagine some person (or whatever your unit of analysis) who could have the combination of qualities described for that cell.

1. Identify your area of interest.

2. Identify the two or more dimensions in your typology and explain why you selected them.

3. Present your typology in a format like that used in the text.

4. Explain your typology and show how it could be used in a research study.

EXERCISE 6.3

Name _____

Select an article from the social science journals available online through your library. Or consult the *Electronic Journal of Sociology* (www.sociology.org) or the *Journal of Criminal Justice and Popular Culture* (www.albany.edu/scj/jcjpc). Your instructor may have other advice. Be sure to select an article that includes an index or scale. Include the citation for your article.

1. Describe the index or scale in your article.

2. Assess the index or scale in terms of the criteria noted in the text.

EXERCISE 6.4

Name _____

The chapter reviews four basic steps in constructing an index: (1) selecting possible items, (2) examining their empirical relationships, (3) combining some items into an index, and (4) validating the index. You are to apply these steps to three items in the General Social Survey codebook (see Appendix 1) that measure attitudes toward abortion. The items are ABRAPE, ABHLTH, and ABDEFECT. Your instructor will advise you to complete all or only some of these steps.

1. Regarding item selection, assess these items in terms of face validity, unidimensionality, and variance. To assess variance, you will have to do a frequency distribution on the items. Use the "valid percent" column for your analyses because this column excludes missing values.

2. Assess the bivariate relationships among these items by doing the following crosstabulations: ABRAPE with ABHLTH, ABRAPE with ABDEFECT, and ABHLTH with ABDEFECT. Request both column and row percentages. Present the tables below and interpret the results.

(continued)

3. Regarding the multivariate (in this case trivariate) relationships among the items, use ABHLTH as the criterion variable (like Babbie used "basic mechanisms" in Figure 6-4). Your instructor may provide specific instructions on this step.

   For the sake of simplicity, the results should be presented in the following form, although other forms would also be appropriate. Remember that column percentages will not necessarily total 100% because the numbers in your table will only represent those who say "yes" to the ABHLTH item.

"Yes" response on ABHLTH

ABRAPE

		Yes	No
	Yes	__%	__%
ABDEFECT			
	No	__%	__%

   Interpret the results:

(continued)

4. Regarding index scoring, apply the discussion in the text to this example by creating a new variable, ABORT, by summing the three items. Recode the variables first so that "No" = 0 and "Yes" = 1. In SPSS, use the *recode* function to recode the three variables into new variables (you might use ABRAP, ABHLT, and ABDEF as names). Recode 0 into 9 (0 is "missing," so you are just recoding one missing value into another so that the 0 can be used for the "no" value), 1 into 1, and 2 into 0. Hence, 1 will still reflect "yes," but 0 will now reflect "no." The index will reflect the number of items the respondent agreed with, among those who responded to all three items. In SPSS, use the *compute* function by simply adding together your recoded variables. Present a frequency distribution of your index and interpret the results.

5. Regarding index validation via item analysis, validate ABRAPE by using the index ABORT as the independent (column) variable and ABRAPE as the dependent (row) variable. Do a crosstabulation and request column percentages. Present and analyze the results.

(continued)

6. Regarding index validation via external validation, validate the index with political views by running a crosstabulation of POLVIEWS by the index you created, ABORT. Make POLVIEWS the row variable and ABORT the column variable and request column percentages. Present and analyze the results.

EXERCISE 6.5   Name _____

Babbie discusses various alternatives for addressing missing values. One of those is to examine other responses of those who have missing values on a variable. Use SPSS or another data analysis program as indicated by your instructor to see if race and gender differences exist for those who are missing on DIVLAW, the item asking if the respondent feels that divorce should be easier or more difficult to obtain. That is, run crosstabulations of DIVLAW by RACE and SEX. Make DIVLAW the row variable and RACE and SEX the column variables (run two tables), and request column percentages. The data file already excludes the missing values from the results because these values have already been defined as missing. Similarly, the missing values are not noted in the codebook in Appendix 1. There are two missing values, 8 for "don't know" and 9 for "not applicable." Examine only those who responded "don't know." So, first change the coding so that this value, 8, is not defined as missing before running your tables. In SPSS you would do this by first clicking "Data Editor," then "Variable View," then "Missing" for the variable DIVLAW, then delete the 8 from the list of missing values. Summarize the differences by race and sex, if any, for those who responded "don't know" versus those who gave one of the coded responses.

1. Report the two tables below.

2. Analyze the results in terms of who is more likely to have the missing values.

ADDITIONAL INTERNET EXERCISES

1. Visit http://www.surveyz.com, a company that does online research. Click "Survey University" near the top, click "Advanced Survey Building" on the left, and click "Survey Question and Answer (Scale) Types." Read the scale tutorial and write suggestions to the company to improve the tutorial. Please note: do not send your suggestions!

3. Visit http://www.mediajunk.com/public/archives/2002/11/virtual_communities_and_social.html http://mediajunk.com/public/and review the story The author reviews the original scale and suggests some new social distance measures such as: I would include a member of group X in my a) address book, b) someone I text-message, c) someone I would allow in my chat group, etc. What do you think of these modifications? Collect data from some friends using these questions and assess, at least at a rudimentary level, whether the data form a scale.

4. Use your library's online resources to locate a social science article that includes scales or indexes. Note the full citation and assess the adequacy of the scales or indexes in terms of Babbie's chapter.

Chapter 7

The Logic of Sampling

OBJECTIVES

1. Define sampling.

2. Document the historical connection between sampling and political polling.

3. Describe and illustrate each of the following types of nonprobability sampling: reliance on available subjects sampling, purposive (judgmental) sampling, snowball sampling, and quota sampling.

4. Describe the role of informants in nonprobability sampling and provide advice on how to select them.

5. Describe the logic of probability sampling, and include heterogeneity and representativeness in your response.

6. List two advantages of probability sampling over nonprobability sampling.

7. Define an EPSEM sample.

8. Define each of the following terms and explain its role in probability sampling: element, population, study population, random selection, sampling unit, and parameter.

9. Using probability sampling theory, describe the sampling distribution.

10. Explain how to interpret a standard error in terms of the normal distribution.

11. Define sampling error and show how confidence levels and confidence intervals are used in interpreting sampling errors.

12. Present an argument that the survey uses of probability theory are not entirely justified technically.

13. Restate Babbie's sociology point regarding making generalizations from sampling frames to populations.

14. Describe simple random sampling and list two reasons why it is seldom used.

15. Summarize the steps in using a table of random numbers.

16. Describe systematic sampling, and employ the concepts of sampling interval, sampling ratio, and periodicity in your description.

17. Link stratified sampling with the principle of heterogeneity and describe how this strategy is executed.

18. Identify the major advantage of multistage cluster sampling and describe how this procedure is executed.

19. Present guidelines for balancing the number of clusters and the cluster size in multistage cluster sampling.

20. Explain why a researcher might use probability proportionate to size sampling, and explain the logic behind this strategy.

21. Outline the rationale for disproportionate sampling and weighting, and note the dangers in using these strategies.

22. Indicate two ethical issues that may enter into sampling design considerations.

OUTLINE

1. A brief history of sampling
   a. President Alf Landon
   b. President Thomas E. Dewey
   c. Two types of sampling methods

2. Nonprobability sampling
   a. Reliance on available subjects
   b. Purposive or judgmental sampling
   c. Snowball sampling
   d. Quota sampling
   e. Selecting informants

3. The theory and logic of probability sampling
   a. Conscious and unconscious sampling bias
   b. Representativeness and probability of selection
   c. Random selection
   d. Probability theory, sampling distributions, and estimates of sampling error

4. Populations and sampling frames
   a. Review of populations and sampling frames

5. Types of sampling designs
   a. Simple random sampling
   b. Systematic sampling
   c. Stratified sampling
   d. Implicit stratification in systematic sampling
   e. Illustration: Sampling university students

6. Multistage cluster sampling
   a. Multistage designs and sampling error
   b. Stratification in multistage cluster sampling
   c. Probability proportionate to size (PPS) sampling
   d. Disproportionate sampling and weighting

7. Probability sampling in review

8. The ethics of sampling

SUMMARY

Having specified what is to be studied, the next task in the research process is selecting a sample. Sampling affords the social scientist the capability of describing a larger population based on only a selected portion of that population. Two broad types of sampling methods are probability sampling and nonprobability sampling. Some sampling methods are more accurate than others; the early failures and the more recent successes of political pollsters illustrate the refinement of sampling procedures and the increasing reliance on probability methods.

Many research situations preclude the use of probability sampling, especially when it is very difficult or impossible to create lists of the elements to be sample. In these situations, researchers employ nonprobability sampling strategies. Reliance on available subjects is used often, but is a very risky method because it limits the generalizability of the results. Purposive (or judgmental) sampling occurs when a researcher selects a sample based on his or her own knowledge of the population, its elements, or the nature of the research study. Snowball sampling is often used in qualitative field research and is particularly useful when the members of a special population are difficult to locate. It involves collecting data on those members of the target population one is able to find and then asking these respondents to provide information needed to locate other members of that population.

Quota sampling strives to attain representativeness by constructing a matrix representing one or more characteristics of the target population, and then collecting data from persons having the required characteristics of a given cell. But it is often difficult to secure adequate information to create accurate quota frames (the proportions that different cells represent), and bias remains a major problem. As a final nonprobability design, field researchers may sometimes use informants, who are members of the group under study, to provide information about the group itself rather than about themselves. It is best to select informants who adequately represent the group under study, but realize that informants' willingness to respond to outsiders may reflect their marginal status within the group.

Because humans are very heterogeneous, it is important to select samples that adequately represent the population under study. Probability sampling allows the selection of samples not subject to the researcher'sociology biases. A sample is considered representative of the population from which it is selected if the aggregate characteristics of the sample closely approximate those same aggregate characteristics in the population. Hence, probability sampling occurs when every element has an equal chance of being selected (an EPSEM sample). Realize that probability samples rarely represent perfectly the populations from which they are drawn. However, probability samples are more representative than other types of samples, and probability theory permits a statistical estimate of the accuracy or representativeness of a sample.

Probability sampling requires familiarity with several components. An element is that unit *about* which information is collected, and it is typically the unit of analysis of the study. There are two levels to which a sample can be generalized. A population is a theoretically specified aggregation of study elements. A study population is the aggregation of elements from which the sample is actually selected. In random selection, each element has an equal chance of selection independent of any other event in the selection process. Random selection enables us to select elements in such a way that descriptions of those elements portrays accurately the total population. A sampling unit is that element or set of elements considered for selection in some stage of sampling.

Two reasons for using random selection methods are that 1) this procedure serves as a check on conscious or unconscious bias on the part of the researcher, and 2) this procedure is based on probability theory, which yields the basis for estimating population characteristics as well as estimating the accuracy of samples. Probability theory is a branch of mathematics that provides the tools researchers need to devise sampling techniques that yield representative samples and to analyze the results of their sampling statistically. Probability theory provides the basis for estimating the parameters of a population. A parameter is the summary description of a given variable in a population.

The calculation and interpretation of sampling error lies at the root of probability sampling theory and is grounded in the principle of the sampling distribution. The sampling distribution dictates that sample statistics will be normally distributed around the population mean. Probability theory also provides formulas for estimating how closely the sample statistics are clustered around the true population value. This is accomplished through the use of confidence levels (how confident we are that our sample estimate is within a set number of sampling errors of the population value) and confidence intervals (the range between the upper and lower values for a given level of confidence). As a result, we can estimate a population parameter and the expected degree of error on the basis of one sample drawn from a population. The logic of confidence levels and confidence intervals also provides the basis for determining the appropriate sample size for a study. Interestingly, the population size is almost irrelevant for determining the accuracy of sample estimates. Unless a sample represents five percent or more of the population it is drawn from, that proportion is irrelevant.

However, probability theory operates on a number of assumptions that are seldom met in real-life survey situations, such as: an infinitely large population, an infinite number of samples, and sampling with replacement. Probability theory is useful only to the extent that the researcher can actually select a probability sample. Researchers sometimes overestimate the precision of estimates produced by using probability theory.

A sampling frame is the list of elements from which a probability sample is selected. Difficulties in securing an adequate and accurate sampling frame put constraints on the use of probability theory. Remember that findings based on a sample can be taken as representative only of the aggregation of elements that compose the sampling frame. Sampling frames do not always truly include all the elements that their names might imply. All elements in a sampling frame must have equal representativeness in order for the results to be generalized even to the population composing the sampling frame.

Simple random sampling is generally assumed in probability applications. This strategy involves assigning a number to each element and using a table of random numbers to select elements for the sample. But simple random sampling is seldom used because it is not generally feasible and it may not be the most accurate method. Systematic sampling is generally preferred over simple random sampling because of simplicity and because it can be more accurate than simple random sampling. Once the sampling ratio (the proportion of the population) is determined, the researcher simply selects the elements corresponding to the sampling interval (the distance between elements selected), with the first element selected with a table of random numbers. But researchers must be on guard for periodicity, a cyclical pattern that coincides with the sampling interval.

Stratification may be used with both these strategies, and it increases representativeness by first organizing the sampling frame into homogeneous groups reflecting variables that may be related to the variables under study. Such homogeneous groupings reduce sampling error. Implicit stratification occurs when a sampling frame is organized by factors relevant to the study.

Multistage cluster sampling is most useful when no master list exists to provide a sampling frame. The researcher employs multiple sampling units such that groups of elements are sampled at different stages; the element is the final stage. But this design produces higher sampling errors because each stage yields additional sampling error. A general guideline is to maximize the number of clusters selected while decreasing the number of elements selected per cluster, because clusters tend to be homogeneous. Cluster designs may employ either simple random or systematic sampling, with or without stratification at any of the stages.

When clusters are of varying sizes, it is important to vary the procedure by employing probability proportionate to size sampling. In this modification, larger clusters are given a greater chance of being selected, but the same number of elements are still selected from each cluster. Sometimes a researcher may deliberately or inadvertently overrepresent a segment of the population. When this occurs, weighting can be used to correct for the disproportionate sampling.

Ethical considerations in sampling design mean that researchers should point out possible sampling errors, flaws in the sampling frame, nonresponse error, or other factors that may make the results misleading. When doing qualitative research, the researcher should be careful not to portray a search for variation as a study in typicality.

TERMS

1. available subjects sampling
2. cluster sampling
3. confidence interval
4. confidence level
5. disproportionate sampling
6. element
7. EPSEM sample
8. heterogeneity
9. informant
10. nonprobability sampling
11. parameter
12. periodicity
18. random selection
19. representativeness
20. sampling
21. sampling bias
22. sampling distribution
23. sampling error
24. sampling frame
25. sampling interval
26. sampling ratio
27. sampling unit
28. simple random sampling
29. snowball sampling

13. population
14. probability proportionate
     to size sampling
15. probability sampling
16. purposive sampling
17. quota sampling

30. statistic
31. stratification
32. study population
33. systematic sampling
34. weighting

MATCHING

_____   1. Another term for probability samples that employ the equal probability of selection method.

_____   2. The theoretically specified aggregation of survey elements to which results obtained from the sample are generalized.

_____   3. A sampling strategy that involves asking members of a special population for information needed to locate others in that population.

_____   4. The summary description of a given variable in a survey sample.

_____   5. The sampling method that is assumed in the statistical computations of social research.

_____   6. A probability method of sampling in which every kth element in the total list is chosen for the sample.

_____   7. A probability method of sampling in which the researcher ensures that appropriate numbers of elements are drawn from homogeneous subsets of the population.

_____   8. Stopping people at a street corner is an example of this type of nonprobability sampling.

_____   9.   That unit about which information is collected and that provides the basis of analysis.

10.   A member of a group who can talk directly about the group as such.
_____

TRUE-FALSE QUESTIONS

T   F   1.   Purposive or judgmental sampling is a type of random sampling.

T   F   2.   Quota sampling begins with a matrix describing the characteristics of the target population.

T   F   3.   A sample in which all members of the population have an equal chance of being selected is known as an EPSEM sample.

T   F   4.   A population is generally greater than a study population.

T   F   5.   A statistic is the summary description of a given variable in a population.

T   F   6.   The standard deviation is the degree of error to be expected for a given sample design.

T   F   7.   A sampling frame is the list of elements from which a probability sample is selected.

T   F   8.   Periodicity is particularly problematic for cluster samples.

T   F   9.   Snowball sampling is a form of probability sampling.

T   F   10.   It is legitimate to sometimes give some cases more weight than others.

REVIEW QUESTIONS

1. The purpose of sampling is to
    a. enhance reliability of measurements.
    b. enhance validity of measurements.
    c. select a set of people with a range of characteristics.
    d. select a set of people who have the same characteristics as the group from which they are taken.
    e. extend the quality of conceptualization and measurement.

2. The failure of several political polls to predict election victories accurately prior to 1950 was due *primarily* to
    a. samples that were too small.
    b. failure to use probability designs.
    c. use of wrong nonprobability designs.
    d. poor timing.
    e. faulty questionnaire construction.

3. Professor Mallory determines the distribution of students on five variables in her target population. She then selects students who fill the pre-established proportions of people in each combination of variables. Which strategy did she use?
    a. available subjects
    b. purposive
    c. stratified
    d. quota
    e. weighting

4. Professor Hilgard wanted to study drug users and began his study by contacting a few drug users that a friend of his knew. He then asked each of these people for the names of other drug users that might be interested in being a part of his study. Which sampling strategy did he use?
    a. available subjects
    b. snowball
    c. quota
    d. purposive
    e. selecting informants

5. The *key* feature of probability sampling is that
   a. large samples are taken.
   b. a large proportion of the larger population is selected.
   c. a sample of individuals from a population must contain essentially the same variations that exist in the population.
   d. sampling error is eliminated.
   e. people are chosen systematically.

6. Professor Rosenberg takes a random sample of students from five Michigan public universities and colleges. He then phrases his report in terms of "all Michigan college students." To what group of elements is he generalizing?
   a. parameter
   b. population
   c. study population
   d. sampling unit
   e. sampling frame

7. Professor Dorfmander did a study on class level and dating at Alexander University. He used the student directory and selected every 15th student. He then mailed questionnaires to them with two follow-ups. He contacted the registrar's office to find out how many students were not listed in the directory but did not sample them. Which one of the following is the sampling frame?
   a. the students on the registrar's list
   b. students both in the directory and not in the directory
   c. the student directory
   d. Alexander University
   e. sampling

8. Inadequate sampling frames yield problems in establishing
   a. generalizability.
   b. theoretical relevance.
   c. accurate hypotheses.
   d. adequate measurements.
   e. all of the above.

9. The principle of sampling distribution suggests that
    a. one sample is as good as another.
    b. many samples of the same population will yield less error.
    c. many samples of the same population will yield values evenly spread out across the confidence interval.
    d. many samples of the same population will yield statistics that will fall around the population value in a known way.
    e. many samples of the same population will yield a lower sampling error than just one sample.

10. Which one of the following distributions of whites and nonwhites in a sample would reflect the most sampling error?
    a. 90-10
    b. 80-20
    c. 70-30
    d. 60-40
    e. 50-50

11. Professor Gherkey calculated a sampling error of 2 in her study on the proportion of students who voted in the last election. She found that 45 percent voted. What is the 95 percent confidence interval?
    a. 41-49
    b. 43-47
    c. 39-51
    d. 44-46
    e. 42-48

12. Professor Andrea did a study on the religious behaviors in one Reformed church. She numbered the members and programmed the computer to generate a set of random numbers and print out address labels. Which type of sampling did she use?
    a. simple random
    b. systematic
    c. stratified
    d. cluster
    e. purposive

13. Professor Connally plans to select a systematic sample of 500 students from a list of 2,000. The sampling ratio would be
    a. 4.
    b. 10.
    c. 8.
    d. one-fourth
    e. none of the above.

14. Professor Stampson wishes to take a sample of university students at one university. His main objective is to make sure that the sample is very representative of the larger population in terms of class level and GPA. Which sampling design would be best?
    a. simple random
    b. systematic
    c. stratified
    d. cluster
    e. snowball

15. Consideration of sampling units is particularly relevant for which type of sampling?
    a. cluster
    b. systematic
    c. stratified
    d. proportional
    e. simple random

16. Professor Gorenpinsky is interested in doing a study comparing male and female members of the National Organization for Women. Because few men belong, he takes three-fourths of the men but only one-thirtieth of the women, selecting members from each segment randomly. Which one of the following *best* describes the principle he used?
    a. Stratification
    b. systematic sampling
    c. grouping
    d. clustering
    e. weighting

17. Lucinda wishes to use a multistage sampling design that will help her take into account that the cities in her sampling frame are vastly different in size. Best to use would be
    a. weighting
    b. periodicity
    c. informants
    d. quota sampling
    e. probability proportionate to size sampling

18. When P is the percent of students voting and Q is the percent of students not voting and n is the sample size, sampling error is calculated by
    a. taking the square root of P times Q divided by n
    b. P times Q divided by n
    c. P plus Q divided by n
    d. taking the square root of P plus Q divided by n
    e. P minus Q divided by n

19. Hortenz took a systematic sample of students in residence halls by taking every 10th student. By coincidence, he ended up with only students in corner rooms. His sampling design reflects
    a. accidental sampling error
    b. random sampling error
    c. confidence intervals
    d. periodicity
    e. weighting

20. Samantha did a study on student involvement in student government by studying only those heavily involved in student government organizations at her school. What design did she employ?
    a. quota
    b. snowball
    c. purposive
    d. simple random
    e. multistage cluster

DISCUSSION QUESTIONS

1. Suppose you were asked by the principal of a local high school to do a survey of the student body. The principal believes it is sufficient to stop students on their way to the library and hand out the questionnaire. You argue for a probability sample instead. State your argument, and give at least two reasons why probability sampling might be preferred. Be sure your analysis incorporates the text discussion on the logic of probability sampling.

2. Discuss the differences between a population, a study population, and a sampling frame. Use examples. Also discuss the cautions that should be applied in generalizing from the sampling frame to the study population and the population.

3. Explain what a sampling distribution is. Why is it an important concept in making inferences from a sample to a population? Link your analysis to the text discussion on probability sampling theory.

4. Determine which of the nonprobability sampling designs is the most representative and explain why.

EXERCISES 7.1 and 7.2

Name _____

The list of sociologists below is taken from a *Guide to Members* published by the American Sociological Association and is to be used in Exercises 7.1 and 7.2.

1.	Lauren Aaronson	37.	Donald Adamchak
2.	Andrew Abbott	38.	Anne Adams
3.	James Abbott	39.	Ben Adams
4.	Kimberly Abbott	40.	Bert Adams
5.	Joan Abbott-Chapman	41.	Darrell Adams
6.	Felix Abdala	42.	Douglas James Adams
7.	Saleha Abedin	43.	Janet Adams
8.	Thomas Abel	44.	Joanne Adams
9.	Anthony Abela	45.	Julia Adams
10.	Ronald Abeles	46.	Laura Adams
11.	Christopher Abells	47.	Michelle Adams
12.	Marjorie Abend-Wein	48.	Randolph Adams
13.	Joel Aberbach	49.	Rebecca Adams
14.	David Aberle	50.	Reed Adams
15.	Pnina Abir-Am	51.	Richard Adams
16.	Joyce Abma	52.	Stacy Adams
17.	Mitchel Abolafia	53.	Christopher Adamson
18.	Jill Abood	54.	Michelle Adato
19.	Mitchell Aboulafia	55.	Lu Ann Aday
20.	Deborah Abowitz	56.	David Aday
21.	Gary Abraham	57.	Michele Adcock
22.	Brant Abrahamson	58.	Donald Addison
23.	Mark Abrahamson	59.	Miriam Adelman
24.	Glen Abrams	60.	Richard Adelman
25.	Paul Abramson	61.	Pamela Adelmann
26.	Jeana Abromeit	62.	Larry Adelmon
27.	Lorien Abroms	63.	Evan Adelson
28.	James Absher	64.	Francis Adeola
29.	Vicki Abt	65.	Maurice Adib
30.	Janet Abu-Lughod	66.	Edward Adlaf

(continued)

31.	Rikki Abzug	67.	Chaim Adler
32.	Joan Acker	68.	Emily Adler
33.	Alan Acock	69.	Gerald Adler
34.	Kazumi Adachi	70.	Glen Adler
35.	Stephen Adair	71.	Leta Adler
36.	Barry Adam	72.	Marina Adler

EXERCISE 7.1

Name _____

Review the box in the text on using a table of random numbers. Using the list of sociologists noted on the previous page and a table of random numbers (see Appendix B in the text), select a simple random sample of ten names.

As you select your sample, you will move through the list of random numbers in the random numbers table. Record below **all** the numbers that you used from the table before you had the numbers necessary for your sample (some will not be used because they are above 72 or are duplicates). Place an asterisk beside the numbers from this original list that you actually used in selecting the names for your sample.

RANDOM NUMBERS USED	SAMPLE OF NAMES SELECTED
____ ____ ____	1.
____ ____ ____	2.
____ ____ ____	3.
____ ____ ____	4.
____ ____ ____	5.
____ ____ ____	6.
____ ____ ____	7.
____ ____ ____	8.
____ ____ ____	9.
____ ____ ____	10.

Explain the procedure you followed to obtain the above sample of names.

EXERCISE 7.2   Name _____

Using the list of sociologists, select a stratified systematic sample of approximately ten names, beginning with a random start. The stratification variable, in this case, is gender.

Reorganize the list of names by gender first. This can most easily be done by simply reorganizing the numbers of the names. If a person's gender is not immediately obvious, place the first non-obvious name in the female category, the second in the male category, etc. Then select your systematic sample across the reorganized (i.e., stratified) list.

1.   Reorganized list:

2.   Please fill in the requested information.

   a.   The sampling interval:

   b.   The random start:

   c.   The list of names selected:

         1.

         2.

         3.

(continued)

4.

5.

6.

7.

8.

9.

10.

11.

NOTE: Correctly drawn systematic samples in this case might result in the selection of 10 or 11 names.

3. Explain the procedure you used in obtaining the above sample.

EXERCISE 7.3

Name _____

You have received a grant to study drug usage among college students in America. You want to be able to make some generalizations about drug usage among *all* college students in America, generalizations about how many students use drugs, the types and frequency of drugs used, and the characteristics of users (such as age and gender). You will be using a stratified cluster sample in which you stratify on at least two variables and sample at least in two stages. There is no single list that you can get that contains the names of all college students in America. So, what sampling units (clusters) can you select at different stages that will ultimately result in a representative sample of *all* college students? Assume that any college can supply you with a list of all of its students. The variables you stratify on should be relevant to your study and should be those for which you have information. For example, you could stratify on grade level, but not on church attendance since the first, but not the second, would be available on the list of names from a college.

Describe the stratified cluster sample you would use. Stratify on at least two variables and explain why you made the decisions you did in term of which variables to use for stratification. Sample at least two stages, including the final sampling unit. Describe the type of sampling procedure you would use at each stage of sampling and explain why you selected the sampling procedures you did at each stage. Be sure to note how many schools, students, etc. that you have in the various stages of sampling. Discuss the limitations of your sample.

EXERCISE 7.4

Name _____

Consult your library's online social science journal resources to locate a social science research article. Select an article that includes a description of the sample used. Apply the following sampling issues and concepts to your article. **Warning:** not all social science research articles are appropriate for this assignment because many do not involve and/or fully describe sampling procedures and sampling error. Your instructor may have additional instructions on what to do in such a situation.

1. Describe briefly the population and study population that the study analyzed.

2. Describe the sampling frame used in the selection of the sample. Does this differ from the population as described in #1? How?

(continued)

3. Describe the type of sampling design used and explain why it was used. Also discuss any particular advantages and/or disadvantages this design yielded.

4. Briefly interpret and assess the standard errors reported, if any.

5. Determine the extent to which the generalizations made are appropriate for the sampling frame used.

ADDITIONAL INTERNET EXERCISES

1. Visit the Annenberg Series online interactive site on sampling at http://www.learner.org/interactives/statistics/ and answer the poll and then follow the directions. The site enables you to review sampling. Summarize the site and what you learned by using the site. Comment on the poll results and the representativeness of the sample.

2. Run through five different sampling scenarios by varying your input values at http://www.custominsight.com/articles/random-sample-calculator.asp. Summarize the impact of changes in the values of the different variables.

3. Work through the tutorial on random digit dialing at http://www.randomizer.org/lesson4.htm and write a brief statement about how the information either supported or conflicted with the information presented in the text.

Part 3

Modes of Observation

Chapter 8

Experiments

OBJECTIVES

1. Give several examples showing that the experimental mode of observation is particularly appropriate for explanatory purposes.

2. Describe and illustrate with an example the three major pairs of components in the classical experiment.

3. Give an example of the double-blind experiment and indicate why such a design would be used.

4. Contrast the following three strategies for selecting subjects: probability sampling, randomization, and matching.

5. Note the feature that the preexperimental designs have in common, and define and develop examples of each of the following three designs: one-shot case study, one-group pretest-posttest design, and static group comparison.

6. Explain how the following factors may threaten internal validity: history, maturation, testing, instrumentation, statistical regression, selection biases, experimental mortality, causal time order, diffusion or imitation of treatments, compensation, compensatory rivalry, and demoralization.

7. Show how the classical experiment handles each of these problems of internal invalidity.

8. Compare the following true experimental designs: classical design, Solomon four-group design, and posttest-only control group design.

9. Show how the true experimental designs address the problem of external validity.

10. Describe how natural experiments occur and give two examples.

11. Examine the strengths and weaknesses of the experimental method.

12. Identify the ethical issues in experimental research.

OUTLINE

1. Topics appropriate to experiments

2. The classical experiment
    a. Independent and dependent variables
    b. Pretesting and posttesting
    c. Experimental and control groups
    d. The double-blind experiment

3. Selecting subjects
    a. Probability sampling
    b. Randomization
    c. Matching
    d. Matching or randomization?

4. Variations on experimental design
    a. Preexperimental research designs
    b. Validity issues in experimental research

SUMMARY

Experimentation involves taking action and observing the consequences of that action. Experiments are particularly well suited to research topics with limited and clearly defined concepts and hypotheses. The experimental model closely parallels the traditional view of science and is particularly appropriate for explanatory research.

The classical experiment involves three major pairs of elements. Such a design examines the effect of an independent variable (the experimental stimulus) on a dependent variable. Both must be clearly defined. Experimental subjects are randomly assigned to an experimental group and a control group, and both are pretested and posttested on the dependent variable. Control groups are used to control for the effects of the experiment itself. One effect that the experimental stimulus may have is known as the Hawthorne effect, in which attention to subjects may affect posttest scores. Double-blind designs are also used to control the effect of the design. In this design neither the subjects nor the experimenters know which is the experimental group and which is the control group.

The main rule for selecting subjects is to maximize the comparability of the experimental and control groups. This can be achieved through three strategies. Probability sampling involves random selection of subjects from a sampling frame for the two groups; it is seldom used. However recruited, subjects may be randomly assigned to either the experimental group or the control group, a technique known as randomization. Third, subjects in the two groups may be matched on variables thought to influence the dependent variable; this is accomplished through the use of a quota matrix. Randomization is the preferred strategy.

Many social scientific experiments employ what are known as preexperimental designs, so called because they lack a legitimate control group and random assignment. The one-shot case study includes a single group of subjects measured on a dependent variable following the occurrence of some experimental stimulus. The one-group pretest-posttest design adds a pretest for the experimental group but still lacks a control group. The static-group comparison includes a "control group," but neither group is pretested, and subjects are not randomly assigned to each.

Preexperimental designs are particularly prone to internal invalidity. Several factors produce such invalidity. Historical events may occur during the course of the experiment that may confound the experimental results. Maturation may also occur, whereby subjects become more tired, bored, or changed in some other respect. The process of testing and retesting can itself influence people's behavior. Or the measures themselves may not be reliable and valid, a problem known as instrumentation.

Statistical regression becomes a problem when subjects are selected for their extreme scores on the dependent variable. Selection biases enter with many of the methods for selecting subjects. Sometimes subjects drop out of an experiment, known as experimental mortality. Occasionally the causal time order is not entirely clear, and sometimes diffusion occurs when subjects in the two groups communicate with each other. Experimenters sometimes decide to compensate the control group, and sometimes subjects in the control group deprived of the experimental stimulus may try to compensate for a missing stimulus by working harder (known as compensatory rivalry). Finally, demoralization may set in as feelings of deprivation among the control group result in their giving up.

The classical design involves an experimental group and a control group, both randomly assigned and pretested, and effectively manages the sources of internal invalidity noted above. Two other true experimental designs are the Solomon four-group design and the posttest-only control group design. The former effectively manages the interaction between the testing and the experimental stimulus by adding two additional groups, neither pretested but both posttested and one receiving the experimental stimulus. The posttest-only control group design involves assigning subjects randomly to an experimental and a control group but pretesting neither. These two designs are better than the others in terms of external validity, also known as generalizability.

In their pure form, experiments are usually conducted in a laboratory setting with a high level of researcher control. But not all social phenomena are subject to this kind of control. As a result, researchers sometimes rely on natural experiments to study effects of events that occur in the real world. Researchers now commonly use the World Wide Web for conducting experiments.

The major advantage of experiments is the clear isolation of the experimental variable and its impact on the dependent variable. Experiments are also easily duplicated, which enhances reliability and validity. The greatest weakness of experiments lies in their artificiality. It is not always clear how well the results of laboratory experiments can be generalized to settings outside the laboratory.

Ethics enter experimental design. Experiments almost always involve some deception, and experi-ments are typically intrusive in the lives of subjects. With both issues, researchers need to balance the potential value of the research against the potential damage to subjects.

TERMS

1. control group
2. demoralization
3. dependent variable
4. double-blind experiment
5. experimental group
6. experimental mortality
7. experimental stimulus
8. external invalidity
9. field experiment
10. Hawthorne effect
11. history
12. independent variable
13. instrumentation

17. natural experiment
18. one-group pretest-posttest design
19. one-shot case study
20. placebo
21. posttesting
22. posttest-only control group design
23. preexperimental designs
24. pretesting
25. quota matrix
26. randomization
27. selection biases
28. Solomon four-group design
29. static-group comparison

14. internal invalidity
15. matching
16. maturation

30. statistical regression
31. testing

MATCHING

_____ 1. In an experiment, the group that receives the experimental stimulus.

_____ 2. The group that does not receive the experimental stimulus.

_____ 3. The process used in matching and based on the most relevant characteristics; half of subjects in each cell go into the experimental group and half into the control group.

_____ 4. A group of designs that deviate from the principles of the classical design.

_____ 5. A type of experiment in which neither the experimenter nor the subjects
know who receives the experimental stimulus.

_____ 6. That which reflects the possibility that the conclusions drawn from the
experimental results may not accurately reflect what has gone on in the experiment itself.

_____ 7. People continually change and these changes can affect the results of an experiment, which is a source of internal invalidity.

_____ 8. A source of internal invalidity in which some events may occur between Pretesting and posttesting and which may confound the experimental results.

_____ 9. Another term for the generalizabilty of experimental findings to the real world.

_____ 10. A source of internal invalidity that involves the process of measurement in pretesting and posttesting.

TRUE-FALSE QUESTIONS

T   F   1.   The control group receives the stimulus in an experiment.

T   F   2.   The double-blind experiment is used to minimize the effects of pretesting.

T   F   3.   Probability sampling is the most popular method for selecting subjects for an experiment.

T   F   4.   The one-group pretest-posttest design provides better evidence than the one-shot case study design that a stimulus produces a particular effect.

T   F   5.   Testing and retesting in an experiment may affect people's behavior and confound the experimental results, a source of internal validity known as instrumentation.

T   F   6.   External invalidity refers to the degree to which experimental results can be generalized to the "real" world.

T   F   7.   The Solomon four-group design is particularly useful to control for the ethical limitations of experiments.

T   F   8.   It is impossible to do experiments out in the "real" world.

T   F   9.   Sometimes it's a good idea to study subjects with extreme scores on the dependent variable, but this could be a source of internal invalidity.

T   F   10.   Experimental mortality refers to the extent to which the researcher drops out of a study and is replaced by a new researcher.

REVIEW QUESTIONS

1. Which one of the following is *false* regarding experimentation? Or are they all true?
   a. The experimental model is closely related to the traditional image of science.
   b. Experimentation is especially appropriate for hypothesis testing.
   c. Experimentation is better suited for explanatory than descriptive research.
   d. Experimentation is especially notable for high researcher control.
   e. All are true.

2. Which one of the following is *not* an aspect of the classical experimental design? Or are they all aspects?
   a. independent and dependent variables
   b. experimental and control groups
   c. high internal validity and high external validity
   d. pretesting and posttesting
   e. all are aspects

3. Professor Stauffer performed an experiment on learning by varying the mode of instruction. She did this for three different grade levels. Which of the following is the experimental stimulus?
   a. mode of instruction
   b. amount learned
   c. anxiety level
   d. grade level
   e. control variable

4. The importance of having a pretest to compare with posttest results lies in
   a. making the sample more random.
   b. comparing the results across pretests.
   c. comparing pretest results with posttest results across groups in the experiment.
   d. making the two groups more equal.
   e. Comparing the pretest with pretests in other experiments to establish generalizability

176

5. An experimenter wanted to see the effects of caffeine intake on the mental state of her subjects. She randomly assigned subjects to the experimental and control groups and administered caffeine-rich soda to one group of subjects and caffeine-free soda to the other group; she then compared the mental states of the members of both groups. This is an example of a
    a. posttest-only control group.
    b. pretest-posttest design.
    c. double-blind experiment.
    d. classical design.
    e. static-group comparison.

6. In the above experiment, caffeine-rich soda and caffeine-free soda, respectively, are the
    a. independent variable and dependent variable.
    b. experimental stimulus and dependent variable.
    c. experimental stimulus and placebo.
    d. placebo and dependent variable.
    e. experimental stimulus and independent variable.

7. Professor Riley performed an experiment in which neither the subjects nor the experimenter knew who got the stimulus. This is an example of which type of experiment?
    a. one-shot case study
    b. classical design
    c. double-blind experiment
    d. static-group comparison
    e. pretest-posttest design

8. Matching refers to
    a. linking subjects in the pretest group with those in the posttest group.
    b. selecting pairs of subjects who are included and not included in an experiment.
    c. selecting similar pairs of subjects and assigning each member randomly to the experimental and control groups.
    d. linking pairs of subjects on the independent variable with those on the dependent variable.
    e. assigning similar pairs of subjects to different settings for the same experiment.

9. Reporter Stocker is involved in testing the effects of watching the news on the emotional health of young children. She decides to use a class of Montessori children and a class of public school elementary children as her "experimental" and "control" groups. This is an example of which type of experiment?
   a. double-blind experiment
   b. static-group comparison
   c. classical design
   d. posttest-only control group design
   e. one-shot case study

10. Professor Flannerly performed an experiment on the effects of interpersonal attraction on conflict orientation, but he used different measures of conflict orientation at the pretest and posttest. Which problem of internal validity does this example reflect?
    a. selection bias
    b. testing
    c. experimental mortality
    d. maturation
    e. instrumentation

11. Elmer, a subject in Professor Jencken's experiment testing the effects of certain films on a person's emotional state, has just undergone a break-up with his girlfriend. He continues with the experiment, however. Which one of the following threats to internal validity is reflected in this example?
    a. history
    b. maturation
    c. selection biases
    d. statistical regression
    e. experimental mortality

12. Professor Holsingley conducted an experiment on the effects of lawyer volunteers on parole violations. Those parolees with access to lawyer volunteers had fewer parole violations than those without, but the control group parolees felt left out and committed more crimes as a result. Which problem of invalidity does this example reflect?
    a. regression
    b. statistical regression
    c. instrumentation
    d. demoralization
    e. selection biases

13. External invalidity refers to problems with
    a. randomization into experimental and control groups.
    b. generalizability.
    c. assessing ethical violations.
    d. assessing posttest effects.
    e. selection.

14. What is the basic difference between the classical design and the Solomon four-group design?
    a. There is no difference.
    b. More time elapses between the stimulus and the second observation in the Solomon four-group design.
    c. The Solomon four-group design has randomization.
    d. The Solomon four-group design repeats the classical design but adds groups that are not pretested.
    e. The Solomon four-group design repeats the classical design but adds groups that are not posttested.

15. Natural experiments are most likely to resemble which one of the following designs?
    a. static-group comparison
    b. classical
    c. Solomon four-group
    d. one-group pretest-posttest
    e. posttest-only control group design

16. What two issues are the most relevant ethical issues in experiments?
    a. deception and intrusion into subjects' lives
    b. intrusion into subjects lives and measurement difficulties
    c. measurement difficulties and pretesting
    d. pretesting and internal invalidity
    e. internal invalidity and deception

17. Yellow Bird is doing an experiment on how social interaction affects prejudice in her community. After the experiment has begun but before it has concluded, someone paints the letters KKK on a couple cars in the community, and outrage follows. Which source of internal invalidity is best demonstrated?
    a. history
    b. instrumentation
    c. statistical regression
    d. maturation
    e. selection biases

18. Juanita did an experiment on the impact of TV watching on students' grades. She was particularly interested in the impact of TV watching on students with high GPAs, so she selected a large proportion of subjects with very high GPAs. Which source of internal invalidity is best demonstrated?
    a. history
    b. instrumentation
    c. statistical regression
    d. maturation
    e. selection biases

19. The major advantage of the experiment is
    a. the high external validity.
    b. the clear isolation of the experimental variable and its impact on the dependent variable.
    c. the presence of few ethical problems.
    d. their natural setting.
    e. their use of large samples.

20. The greatest weakness of experiments is
    a. their small sample sizes.
    b. their limited internal validity.
    c. their connection to the real world.
    d. the reliance on only one or a few independent variables.
    e. their artificiality.

DISCUSSION QUESTIONS

1. Describe the classical experiment. Explain the importance of each of the three major pairs of components.

2. Compare and contrast the three basic strategies for making the subjects in the experimental and control groups as similar as possible. Which one is best? Why? Which technique is actually used most frequently? Why?

3. Differentiate internal from external invalidity. Describe at least two factors contributing to each. Which one of the two—internal or external invalidity—is more serious? Why?

4. Which features of the classical experimental design would be most difficult to incorporate in a natural experiment? Why?

EXERCISE 8.1

Name _____

Your assignment is to design a small-group laboratory experiment that would test the following hypothesis using the pretest-posttest control-group (classical) design:

Men who work with women who appear clearly more effective than themselves in finding a solution to a problem-solving task will have a higher regard for women than men who work in a group without women on the same task.

Note: This experiment involves *only* men as subjects. The women are simply confederates and are not part of the basic design. Also, be sure to devise a laboratory experiment.

1. Describe exactly where and how you would select the men for the study. How many men would you select? Why?

2. Describe the experimental and control groups appropriate to the experiment. Describe how you would assign the men to the experimental and control groups.

(continued)

3. Describe a problem-solving task that might be appropriate for this experiment. That is, develop a problem-solving task that would offer the male subjects the experience of seeing women do better than themselves (the experimental group) or the experience of working on the problem-solving task without the more competent women or without any women (the control group). Again, remember that the women are confederates and are not part of the experimental or control groups.

4. Describe the means by which you would test the attitudes of the men in the study regarding their views about women's analytical abilities (in general, not necessarily regarding the women confederates) *prior* to the experiment. This is the *pretest* of attitudes. Provide three sample items.

(continued)

5. Describe how you would make sure that the confederate women in the experimental group would be clearly more effective in solving the problem than the men.

6. Describe the means by which you would test the attitudes of the men in the study regarding their views about women's analytical abilities (in general, not necessarily regarding the women confederates) *after* the experiment. This is the *posttest* of attitudes. Provide three sample items.

7. Describe two potential sources of internal invalidity that your design reduces or avoids. Explain why or how your design does this.

(continued)

8. Describe one potential source of external invalidity that your design reduces or avoids. Explain why or how your design does this. Remember, external invalidity reflects the generalizability of your results and hence pertains mostly to how you selected your men.

9. Make up some "results" that you might get in such an experiment and interpret them. Be sure to indicate whether the "results" tend to confirm or disconfirm the hypothesis.

EXERCISE 8.2

Name _____

Your assignment is to conduct an experiment on your friends and acquaintances to examine their sincerity in asking and answering the question, "How are you?" Identify the hypothesis you will test (see the table in #6 first). Your experiment will involve three groups of five subjects each: a control group and two experimental groups. Subjects will be placed in the groups as follows.

Control Group: As you meet friends over the next few days, greet five of them by saying, "Hi, how are you?" Record exactly how they answer.

Experimental Group 1: For five other friends, allow them to ask how you are first, and reply by saying, "Great, how are you?" Record exactly how they answer.

Experimental Group 2: For five other friends, allow them to ask how you are first, and reply by saying, "Terrible, how are you?" Record exactly how they answer.

In assigning friends to the two experimental groups, be sure to alternate "great" and "terrible," saying "great" to one and "terrible" to the next, and so forth. Get people into the control group whenever you are able to beat them to the punch.

1.   Describe your hypothesis:

2.   Responses given by subjects in the control group:

   1.

   2.

   3.

   4.

   5.

(continued)

3.   Responses given by subjects in Experimental Group 1 ("Great, how are you?"):

1.

2.

3.

4.

5.

4.   Responses given by subjects in Experimental Group 2 ("Terrible, how are you?"):

1.

2.

3.

4.

5.

(continued)

5.   Score each of the responses as "positive," "neutral," or "negative." For example, a subject who answered "great" would be scored as "positive," one who answered "lousy" would be scored as "negative," and one who answered "so-so" would be scored as "neutral." Your task is to figure out the proper scoring for other kinds of responses. Go back to your answers above and write "positive," "neutral," or "negative" beside each of the responses. *Describe any difficulties you experienced in coding.*

6.   Complete the table below, entering the percentages of "positive," "neutral," and "negative" responses received from each of the three groups.

	Control Group	Experimental Group 1	Experimental Group 2
Percent Positive	%	%	%
	_____	_____	_____
Percent Neutral	%	%	%
	_____	_____	_____
Percent Negative	%	%	%
	_____	_____	_____
	100%	100%	100%
	(5)	(5)	(5)

7.   Provide a brief interpretation of your research findings in light of your hypothesis.

(continued)

8.   Describe the success or failure you experienced in assigning people randomly to the groups.

9.   Assess this design (not the results themselves) in terms of three sources of internal invalidity.

10.   Assess this design (not the results themselves) in terms of external invalidity. (Remember: external invalidity refers to generalizability.)

EXERCISE 8.3

Name _____

You have been asked to develop a study of the effects of watching *Sesame Street* on children's gender-role orientations, intelligence, or sociability (select one). Describe how you would design the study using one preexperimental design: one-shot case study, one-group pretest-posttest design, or static-group comparison—select one. Describe how you would design the study using one true experimental design: pretest-posttest control-group (classical) design, Solomon four-group design, or posttest-only control-group design—select one. Note the major advantage and weakness of each design in this particular study. You might consider the various sources of internal invalidity Babbie mentions as possible weaknesses, although which ones are appropriate varies according to the study design.

1.   Dependent variable you selected:

2.   Preexperimental design you selected:

3.   Describe study using this design:

4.   Advantage of preexperimental design:

(continued)

5.   Weakness of preexperimental design:

6.   True experimental design you selected:

7.   Describe study using this design:

8.   Advantage of true experimental design:

9.   Weakness of true experimental design:

EXERCISE 8.4

Name _____

As a contrived type of natural experiment, we may investigate if owning a gun (the "stimulus") affects propensity toward violence. Use SPSS or another data analysis program as indicated by your instructor to test this hypothesis by running a crosstabulation of POLHITOK (it is OK for police to strike a citizen) by OWNGUN (own an gun or not). Make POLHITOK the row variable and OWNGUN the column variable, and request column percentages.

1.   Report your table below.

2.   Analyze the results to determine if the hypothesis is correct.

ADDITIONAL INTERNET EXERCISES

1. Visit http://psych.athabascau.ca/html/Validity/concept.shtml, which is a tutorial site on internal validity. This site covers via definition, example, and a "test" the following threats to internal validity: 1) selection, 2) history, 3) maturation, 4) repeated testing, 5) instrumentation, 6) regression to the mean, 7) experimental mortality, 8) selection-maturation interaction, and 9) experimenter bias. Complete Part 1 by going through the explanations of the nine threats to internal validity. Then complete Part 2 (visible at the end of Part 1) and indicate your agreement or disagreement with at least ten of the scenarios presented. If you disagree with an answer and believe another threat to internal validity is crucial, explain your disagreement.

2. Visit http://www.une.edu.au/WebStat/unit_materials/c2_research_designhttp://www.une.edu.au/WebStat/to review experimental and quasi-experimental designs. Click "Enter," click "Unit Materials," and click "C2–Research Design." Read the materials under each subheading. What did the site contain that differed from the text's presentation of the material?

3. Visit the web site of the Stanford prison experiment (http://www.prisonexp.org) and use that experiment to illustrate several points made in the chapter.

Chapter 9

Survey Research

OBJECTIVES

1. Illustrate how surveys may be used for descriptive, explanatory, and exploratory purposes.

2. Describe how surveys are sometimes misused.

3. Differentiate questions from statements by definition and example.

4. Outline the conditions under which open-ended and closed-ended questions are used.

5. List and illustrate several guidelines for asking effective questions.

6. Explain why social desirability is a problem in asking questions.

7. List three guidelines for good questionnaire format.

8. Describe the role of contingency questions and list the principles for their use.

9. Describe the role of matrix questions and list the principles for their use.

10. Explain why the order in which questions are asked is important, and describe how this principle is differentially applied in questionnaires and interviews.

11. List three principles for providing instructions for respondents of surveys.

12. List and compare three methods of administering survey questionnaires.

13. List three methods for distributing self-administered questionnaires.

14. List three principles for mail distribution and return of questionnaires.

15. Present an argument for monitoring returns, and show how this can be done with the return rate graph.

16. List three principles regarding follow-up mailings.

17. Present four advantages of interviews over questionnaires.

18. Restate the five general rules for successful interviewing.

19. Discuss the role of specifications in training interviewers.

20. Describe several elements of interviewer training.

21. List the advantages and problems with telephone surveys.

22. Show how computer-assisted telephone interviewing overcomes some of the weaknesses of the telephone survey.

23. Describe several variations for using computers for administering self-administered questionnaires.

24. Describe the advantages of online polling and offer some advice for successful online polling.

25. Contrast self-administered questionnaires, face-to-face interviews, and telephone interviews, and describe when each is most appropriate.

26. Assess the strengths and weaknesses of survey design.

27. Give two examples of secondary analysis and/or data archives, and summarize the advantages and disadvantages of this approach.

28. Identify relevant ethical concerns in survey research.

OUTLINE

1. Topics appropriate for survey research

2. Guidelines for asking questions
    a. Choose appropriate question forms
    b. Make items clear
    c. Avoid double-barreled questions
    d. Respondents must be competent to answer
    e. Respondents must be willing to answer
    f. Questions should be relevant
    g. Short items are best
    h. Avoid negative items
    i. Avoid biased items and terms

3. Questionnaire construction
    a. General questionnaire format
    b. Formats for respondents
    c. Contingency questions
    d. Matrix questions
    e. Ordering items in a questionnaire
    f. Questionnaire instructions
    g. Pretesting the questionnaire
    h. A sample questionnaire

4. Self-administered questionnaires
    a. Mail distribution and return
    b. Monitoring returns
    c. Follow-up mailings
    d. Response rates
    e. A case study

5. Interview surveys
    a. The role of the survey interviewer
    b. General guidelines for survey interviewing
    c. Coordination and control

6. Telephone surveys
    a. Computer-assisted telephone interviewing (CATI)
    b. Response rates in interview surveys

7. Online surveys

8. Comparison of the different survey methods

9. Strengths and weaknesses of survey research

10. Secondary analysis

11. Ethics and survey research

SUMMARY

Survey research is the most frequently used mode of gathering data in the social sciences. It consists of selecting a sample of respondents and administering a questionnaire or an interview to them. Surveys may be used for descriptive, explanatory, and exploratory purposes. The unit of analysis is usually the individual, but surveys can be used for other units of analysis as well. Survey research is particularly well suited for studying attitudes and orientations in a large population. Surveys are increasingly used for less legitimate purposes than social research, including marketing surveys and political surveys such as "push polls."

Most concepts are measured through questions and statements. Questions may take one of two forms. Closed-ended questions are easily processed, but sometimes overlook alternative answer categories. Open-ended questions afford a wider range of responses, but they are more difficult to process and may yield irrelevant answers.

Questions should also be relevant to the respondent, very clear, and as short as possible. Respondents should be capable of answering the questions and should be willing to answer the questions. It is best to avoid negative items as well as biased items and terms. Combining two (or more) questions into one is known as a double-barreled question and produces both confusion for the respondent and ambiguity for the researcher. The researcher should also avoid using questions that promote social desirability—answers that make respondents look good. In the final analysis, the overriding guideline is that questions should reflect the purpose of the study.

Questionnaire format and appearance are critical. Questionnaires that are well organized, uncluttered, and attractive reduce the likelihood that respondents will overlook or ignore items or dispose of the questionnaire. The best format for questions is the use of boxes adequately spaced. Contingency questions can help prevent respondents from being confused and annoyed–questions that are relevant to only some of the respondents are identified. A series of questions with the same set of answer categories can be effectively presented through a matrix format, which saves space and time. But occasionally respondents begin marking the same answers without carefully reading the items, a problem known as a response set.

The order of question presentation is also critical because answers to one question can affect answers to later questions. Also, it is usually best to begin questionnaires with the most interesting set of questions to generate interest, although it is generally best to begin interviews with demographic questions to enhance rapport. Clear instructions at the beginning of a questionnaire, as well as at the beginning of each section, are mandatory. Special situations will require special instructions. Finally, it is important to pretest your questionnaire to help reduce error. Although it is not essential that pretest participants reflect a representative sample, they should use people for whom the questionnaire is at least relevant.

Self-administered questionnaires are generally executed by mail, although sometimes they are administered to a group of respondents, and sometimes they are delivered and picked up at a later time. Self-mailing questionnaires require no return envelopes, making it easier to return the instrument. Questionnaires can be sent via first-class postage or bulk rate and returned via postage stamps or business-reply permits. When each option would be used is determined largely by the budget and the expected return rate.

Monitoring the returns of questionnaires is an important part of the study and can be accomplished with a return rate graph plotting the number of questionnaires returned each day. Such a graph can provide clues about when follow-up mailings should occur, and it may be useful in estimating nonresponse bias. Follow-up mailings may contain only a letter of additional encouragement or may contain a new copy of the questionnaire as well. Follow-up mailings significantly increase return rates; two follow-ups mailed two or three weeks apart are best.

Interviews provide a number of advantages over the self-administered questionnaire. They attain higher response rates, decrease the number of "don't know" and "no answer" responses, help correct confusing items, and provide the opportunity to observe the social situation as well as to ask questions.

It is critical that the interviewer serve as a neutral conduit through which questions and answers are transmitted. Interviewers should have a pleasant demeanor and dress in a fashion similar to the people being interviewed. Familiarity with the instrument will lessen the time required and minimize problems with individual items. Interviewers should follow the question wording exactly and record responses exactly. One particular skill required is probing, whereby the interviewer redirects inappropriate answers while maintaining the neutral role. When multiple interviewers are employed, the researcher should thoroughly train and supervise the interviewers. Researchers should also prepare specifications, which are explanatory and clarifying comments about handling difficult or confusing situations that may arise with a specific item.

Telephone surveys save money and time, sometimes enhance willingness to give socially disapproved answers, and yield greater control when several interviewers are employed. They also alleviate personal safety concerns. But they come with several problems. They have a bad reputation, for example, and people can hang up easily before the interview is completed. Unlisted numbers can be problem unless the researcher uses random-digit dialing. The rise in cell phone usage is also a concern because cell phone numbers are generally excluded in phone surveys. And some people, generally younger adults, use cell phones exclusively. A proliferation of bogus "surveys" used for telemarketing has reduced people's willingness to participate. Answering machines and voicemail may also create problems as people screen their calls.

Computer-assisted telephone interviewing is frequently used to enhance the reliability and validity of the results. Computer-assisted survey methods are also increasingly being used for in-person interviews as well as for analyzing data as the data are gathered. Recent innovations in computer technology provide several options for computer administered questionnaires. Online polling offers advantages over other forms but comes with some disadvantages, particularly concerning representativeness. But online researchers need to be consistent in wording, use plain language, offer a summary of results, be sensitive to time of emailing the survey, be aware of technical limitations, test incentives, and limit studies to 15 minutes. Online surveys seem to have response rates similar to mail surveys.

Self-administered questionnaires and interviews each have advantages and disadvantages. Questionnaires are generally cheaper and quicker and are more appropriate in dealing with more sensitive issues. Interviews help reduce incomplete answers, are more effective in dealing with complicated issues, enable a researcher to conduct a survey based only on a sample of addresses, and provide the opportunity to make important observations apart from the responses to the items. Interviews generally yield higher response rates.

The survey design itself has strengths and weaknesses compared to other designs. It is particularly appropriate for describing the characteristics of large populations, thereby making large samples feasible. In some respects, surveys are quite flexible, and the use of standardized questions enhances reliability. But these advantages come at some cost. Standardization often results in overlooking other appropriate responses and may generate inflexibility in modifying questions. Surveys seldom analyze the context of social life. They are frequently labeled as artificial because the topic of study may not be amenable to measurement through questionnaires, and the act of studying that topic may affect the results. In short, survey research is weak on validity and strong on reliability.

Researchers with a need for survey data cannot always afford to conduct large-scale surveys themselves. Hence many researchers employ secondary analysis of data gathered for some other purpose by someone else. Such research has experienced increased popularity due to the establishment of data archives, research centers that collect and distribute many different types of data sets. Such analysis can be done quite rapidly and inexpensively, particularly with those that are online. The key advantages of secondary analysis include the low cost and the ability to use data gathered by well known researchers and data collection agencies. The key problem with secondary analysis is validity; when one researcher collects data for a given purpose, other researchers have no assurance that such data will be appropriate for their research interests. Comparability of measures is a particular problem. Regarding ethics, surveys often ask for private information (which should remain confidential), and asking questions may cause psychological discomfort (which should be minimized).

TERMS

1. bias
2. closed-ended question
3. Computer Assisted Telephone Interviewing
4. contingency question
5. data archives
6. double-barreled quesion
7. interview
8. matrix questions
9. meta-analysis
10. open-ended question
11. probe
12. questionnaire
13. respondent
14. response rate
15. response set
16. return rate graph
17. secondary analysis
18. self-administered questionnaire
19. social desirability
20. specifications
21. telephone surveys

MATCHING

_____ 1. When people answer survey questions through a filter of what will make them look good.

_____ 2. The percentage of those sent a survey who return them.

_____ 3. Collections of data sets that are available to researchers.

_____ 4. A method that helps researchers monitor the flow of questionnaires returned by respondents.

_____ 5. A question that asks respondents for a single answer to a question that actually has multiple parts.

_____ 6. Explanatory and clarifying comments about the handling of difficult or confusing situations that may occur in an interview situation.

_____ 7. A method used by interviewers to obtain elaboration of respondents' answers to open-ended questions.

_____ 8. A method of research in which researchers analyze data that another researcher has collected.

_____ 9. A type of question in which the respondent is asked to select an answer from among a list provided by the researcher.

_____ 10. A system for telephone interviewing that involves presentation of questions on a computer screen and that also prepares data for analysis.

TRUE-FALSE QUESTIONS

T   F   1. Exhaustive questions include all the possible responses that might be expected.

T   F   2. Double-barreled questions are a good thing to use because they allow the researcher to ask two questions in one.

T   F   3. It is a good idea to include negatively worded questions so that the researcher can determine if respondents are taking the questions seriously.

T   F   4. Open blanks are the best format to use for questions.

T   F   5. Contingency questions are used when a researcher needs to ask additional questions only of a segment of the respondents.

T   F   6. Monitoring returns is useful only for deciding when to send out follow-up mailings.

T   F   7. Probes are used in a pretest of an interview to determine reliability and validity.

T   F   8. Surveys are particularly useful for describing in detail the nature of small segments of the population.

T   F   9. Telephone surveys can allow greater control over data collection if several interviewers are engaged in the project.

T   F   10. As a rule, interviewers should dress in a fashion similar to that of the people they will be interviewing.

REVIEW QUESTIONS

1. The *major* problem with closed-ended questions is
    a.   the number of items in the questionnaire.
    b.   the number of items per dimension.
    c.   the order of the answers.
    d.   the structuring of the responses.
    e.   response set.

2. Read the following item and then answer the question that follows: "How satisfied are you with your working conditions and wages?"
    a.   very satisfied
    b.   somewhat satisfied
    c.   somewhat dissatisfied
    d.   very dissatisfied

3. What is the *major* weakness with this item?
    a.   It is double-barreled.
    b.   It is not clear.
    c.   It is too long.
    d.   It is not relevant.
    e.   Respondents may not be competent to answer it.

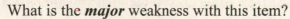

4. Read the following item and then answer the question that follows: "How old were you when you first got in a fight with someone other than a sibling?"
    a.   two years old or younger
    b.   three to five years old
    c.   six to seven years old
    d.   eight to nine years old
    e.   ten years old or older
    f.   I have never been in a fight

5. What is the *major* weakness with this item?
    a.   It is double-barreled.
    b.   It is biased.
    c.   It is too long.
    d.   It is not relevant.
    e.   Respondents may not be competent to answer it.

4   6.   Read the following item and then answer the question that follows: "The main cause of highway accidents is speeding by truckers. Don't you agree that the use of radar detectors by truckers should be made illegal?"
      a.   Yes
      b.   Not sure
      c.   No

7.   What is the *major* weakness with this item?
      a.   It is socially desirable.
      b.   It is too long.
      c.   It is biased.
      d.   It is double-barreled.
      e.   It is not clear.

5   8.   Read the following item and then answer the question that follows: "How often do you feel that you are basically a kind and loving person?"
      a.   all the time
      b.   usually
      c.   usually not
      d.   never

9.   What is the *major* weakness with this item?
      a.   It is socially desirable.
      b.   It is too long.
      c.   It is not clear.
      d.   It is not mutually exclusive.
      e.   Respondents may not be competent to answer it.

6   10.   Which one of the following is the *best* method for providing space for checking responses? Or are they all equally fine?
      a.   boxes
      b.   parentheses
      c.   brackets
      d.   open blanks for check marks
      e.   all are equally fine

11. Professor Kaled wishes to ask three additional questions only of those respondents who have been active in a political organization in the previous year. *Best* to use would be
    a.   contingency questions.
    b.   matrix questions.
    c.   matched questions.
    d.   separate questionnaires.
    e.   different response sets.

12. Response set is *most* likely to occur in which kinds of questions?
    a.   matrix questions
    b.   contingency questions
    c.   closed-ended questions
    d.   open-ended questions
    e.   interview questions

13. The major problem with online surveys is
    a.   cost.
    b.   privacy.
    c.   length of survey.
    d.   representativeness of respondents.
    e.   level of language used.

14. An important feature of using a return rate graph is to
    a.   reduce sampling error.
    b.   examine history effects.
    c.   examine measurement effects.
    d.   estimate nonresponse biases.
    e.   improve reliability.

15. Interviewers' training sessions should begin with
    a.   discussing specifications.
    b.   discussing probes.
    c.   a description of what the study is about.
    d.   a demonstration interview.
    e.   practice interviews.

16. Which one of the following is *not* an advantage of interviews over questionnaires? Or are they all advantages?
    a.   increased response rates
    b.   safeguard against confusing items
    c.   lower numbers of "don't know" and "no answer" responses
    d.   increased reliability
    e.   all are advantages

17. Interviewers can be helped in dealing with confusing situations regarding a given item through the use of clarifying comments known as
    a.   specifications.
    b.   elaborations.
    c.   matrix questions.
    d.   response set formats.
    e.   conversations.

18. Which one of the following is *false* regarding telephone interviews?
    a.   They are cheaper than in-person interviews.
    b.   They save time over in-person interviews.
    c.   They enhance the safety of the interviewer.
    d.   They make it harder for the respondent to terminate the interview.
    e.   They have a bad reputation.

19. The major problem with secondary analysis pertains to
    a.   theory.
    b.   hypotheses.
    c.   validity.
    d.   sampling.
    e.   empirical generalization.

20. Probes are particularly useful in eliciting responses to which type of question?
    a.   double-barreled
    b.   secondary
    c.   open-ended
    d.   closed-ended
    e.   negative items

21. Which one of the following is correct regarding the desired ordering of items?
    a. begin questionnaires with the most interesting items, begin interviews with demographic data
    b. begin questionnaires with demographic data, begin interviews with the most interesting items
    c. begin questionnaires with the hardest questions, begin interviews with the easiest questions
    d. begin questionnaires with the easiest questions, begin interviews with the hardest questions
    e. it really doesn't matter which items you begin with in either format

22. Follow-up mailings should occur approximately how soon after the first or previous mailing?
    a. about one week
    b. about two or three weeks
    c. about four or five weeks
    d. about six or seven weeks
    e. about eight or nine weeks

23. Which one of the following is the best advice when doing interviews?
    a. follow question wording exactly, but interject your interpretations in recording responses
    b. interject your interpretations in the question wording, but record responses exactly
    c. interject your interpretations in the question wording as well as in recording responses
    d. follow question wording exactly and record responses exactly
    e. first determine the mood of the respondent and then decide how exactly you need to ask questions and record responses

24. The two major ethical concerns in survey research are
    a. honesty and providing incentives.
    b. providing incentives and indicating your source of funding.
    c. indicating your source of funding and maintaining confidentiality.
    d. maintaining confidentiality and a risk of psychological injury.
    e. a risk of psychological injury and honesty.

DISCUSSION QUESTIONS

1. Why is a low response rate of concern in survey research? Why is response bias of concern? Which is a more serious threat to survey results and generalization—low response rates or response biases? Why?

2. Compare and contrast interviews and questionnaires on the several dimensions noted in the chapter.

3. Write a first draft of a brief guide for new interviewers.

4. In this chapter Babbie describes many guidelines for asking questions. Discuss how a researcher could best avoid making the errors he describes. Discuss strategies appropriate before, during, and after the question-writing stage.

NO

EXERCISE 9.1

Name _____

Use the information that follows to construct both types of return rate graphs that Babbie discusses. Then write a short analysis of the results.

A questionnaire was mailed to 1,000 students (both off-campus and on-campus) at Northern Michigan University on March 1. A follow-up was done on March 17. The questionnaires were returned as follows:

Date	# Received	Date	# Received
March 6	2	March 26	no mail
March 7	5	March 27	no mail
March 8	18	March 28	28
March 9	32	March 29	21
March 10	35	March 30	23
March 11	25	March 31	14
March 12	no mail	April 1	12
March 13	no mail	April 2	no mail
March 14	20	April 3	no mail
March 15	19	April 4	10
March 16	20	April 5	11
March 17	14	April 6	9
March 18	13	April 7	6
March 19	no mail	April 8	5
March 20	no mail	April 9	no mail
March 21	16	April 10	no mail
March 22	26	April 11	2
March 23	37	April 12	0
March 24	41		
March 25	30		

1. Create a graph that plots the number of questionnaires returned each day.

(continued)

2. Create a graph that plots the cumulative number over time and percentage of questionnaires returned.

3. Summarize the results of both graphs.

EXERCISE 9.2

Name _____

Each of the questionnaire items below has two or more things wrong with it. Point out at least two defects for each questionnaire item. Rewrite each item to correct the defects. Do not use open-ended questions as a rewrite to fix problems in closed-ended questions.

1. Religion
   Protestant   Jewish
   Catholic   Baptist
   Lutheran   Other

   Defects:

   Rewritten:

2. At what age were you toilet-trained?

   ( ) Before six months old
   ( ) Between six months old and nine months old
   ( ) Between nine months old and one year old
   ( ) Between one year old and one and a half years old
   ( ) Between one and a half years old and two years old
   ( ) Between two years old and three years old
   ( ) Older than three years old
   ( ) Not applicable

   Defects:

   Rewritten:

(continued)

3.　　How much money do you make? $

　　　Defects:

　　　Rewritten:

4.　　Suppose you were in a bookstore and saw a book displayed on a
　　　counter near the door that you wanted very much but could not
　　　afford. Would you steal it?
　　　( ) Yes
　　　( ) No

　　　Defects:

　　　Rewritten:

(continued)

5.   Are you a college student or graduate? If so, why did you decide to go to college?

( ) I had a thirst for more knowledge
( ) I wanted to get a better understanding of the world
( ) I was too lazy to get a job

Defects:

Rewritten:

6.   Do you agree or disagree that the trouble with welfare is that people get too comfortable and don't want to go back to work, so the government should institute some job-training programs for people on welfare and then set a limited amount of time in which they can learn work skills and get a job?
( ) Agree
( ) Disagree

Defects:

Rewritten:

7.   Do you disagree or agree with the President that the United States shouldn't provide a national health care program?
( ) Agree
( ) Disagree

Defects:

Rewritten:

(continued)

8.   Where do you get most or all of your information about current
      events in the nation and the world?
      ( ) Radio
      ( ) Newspapers
      ( ) Magazines
      ( ) Internet

      Defects:

      Rewritten:

9.   Why do you think big cars are a bad thing for America?

      Defects:

      Rewritten:

EXERCISE 9.3

Name _____

On the following pages are ten copies of a short interview schedule. Your assignment is to select and interview ten people: five men and five women (your instructor may provide additional sampling instructions). Do *not* simply interview your friends. Please be sure to actually interview the respondents; do *not* have them fill out their own questionnaires. Do *not* read out loud the specifications (the comments in all capitals in parentheses).

After completing the interviews, write a short report of your experiences below. Discuss the problems you experienced and how you dealt with them. Which types of probes did you employ? With what success? Which aspects of the advice for interviewing found in the chapter proved to be the most useful? Which proved to be the least useful? What would you do differently now? Why? How? Do not discuss the actual results of the survey.

(continued)

Case ID_____   Respondent's gender: ☐ Male ☐ Female

Hello. My name is _____ and I'm conducting a short survey as part of an exercise for my research methods class. Would you be willing to participate in the study? The interview will take less than five minutes. (**ENCOURAGE IF RELUCTANT.**)

We are faced with many problems in this country, none of which can be solved easily or inexpensively. I'm going to name some of these problems, and for each one I'd like you to tell me whether you think we're spending too much money on it, too little money, or about the right amount. Are we spending too much, too little, or about the right amount on . . . (**READ EACH ITEM AND CHECK THE APPROPRIATE CATEGORY; CODE DK FOR "DON'T KNOW" OR NA FOR "NO ANSWER" IF APPROPRIATE.**)

	Too little	About right	Too much	DK	NA
1. Welfare?	☐	☐	☐	☐	☐
2. Halting the rising crime rate?	☐	☐	☐	☐	☐
3. Improving the nation's education system?	☐	☐	☐	☐	☐
4. Foreign aid?	☐	☐	☐	☐	☐

I am going to name some institutions in this country. As far as the *PEOPLE RUNNING* these institutions are concerned, would you say that you have a great deal of confidence, only some confidence, or hardly any confidence at all in them? (**CODE FOR EACH ITEM; CODE DK FOR "DON'T KNOW" OR NA FOR "NO ANSWER" IF APPROPRIATE.**)

	Great Deal	Only Some	Hardly Any	DK	NA
5. Education	☐	☐	☐	☐	☐
6. Press	☐	☐	☐	☐	☐
7. Congress	☐	☐	☐	☐	☐

8. Could you tell me, in your own words, how you feel about capital punishment?

That completes the interview. Thank you for your cooperation.

Case ID_____   Respondent's gender: ☐ Male ☐ Female

Hello. My name is _____ and I'm conducting a short survey as part of an exercise for my research methods class. Would you be willing to participate in the study? The interview will take less than five minutes. (**ENCOURAGE IF RELUCTANT**.)

We are faced with many problems in this country, none of which can be solved easily or inexpensively. I'm going to name some of these problems, and for each one I'd like you to tell me whether you think we're spending too much money on it, too little money, or about the right amount. Are we spending too much, too little, or about the right amount on . . . (**READ EACH ITEM AND CHECK THE APPROPRIATE CATEGORY; CODE DK FOR "DON'T KNOW" OR NA FOR "NO ANSWER" IF APPROPRIATE**.)

	Too little	About right	Too much	DK	NA
1. Welfare?	☐	☐	☐	☐	☐
2. Halting the rising crime rate?	☐	☐	☐	☐	☐
3. Improving the nation's education system?	☐	☐	☐	☐	☐
4. Foreign aid?	☐	☐	☐	☐	☐

I am going to name some institutions in this country. As far as the *PEOPLE RUNNING* these institutions are concerned, would you say that you have a great deal of confidence, only some confidence, or hardly any confidence at all in them? (**CODE FOR EACH ITEM; CODE DK FOR "DON'T KNOW" OR NA FOR "NO ANSWER" IF APPROPRIATE**.)

	Great Deal	Only Some	Hardly Any	DK	NA
5. Education	☐	☐	☐	☐	☐
6. Press	☐	☐	☐	☐	☐
7. Congress	☐	☐	☐	☐	☐

8. Could you tell me, in your own words, how you feel about capital punishment?

That completes the interview. Thank you for your cooperation.

Case ID_____   Respondent's gender: ☐ Male ☐ Female

Hello. My name is _____ and I'm conducting a short survey as part of an exercise for my research methods class. Would you be willing to participate in the study? The interview will take less than five minutes. (**ENCOURAGE IF RELUCTANT**.)

We are faced with many problems in this country, none of which can be solved easily or inexpensively. I'm going to name some of these problems, and for each one I'd like you to tell me whether you think we're spending too much money on it, too little money, or about the right amount. Are we spending too much, too little, or about the right amount on . . . (**READ EACH ITEM AND CHECK THE APPROPRIATE CATEGORY; CODE DK FOR "DON'T KNOW" OR NA FOR "NO ANSWER" IF APPROPRIATE**.)

	Too little	About right	Too much	DK	NA
1. Welfare?	☐	☐	☐	☐	☐
2. Halting the rising crime rate?	☐	☐	☐	☐	☐
3. Improving the nation's education system?	☐	☐	☐	☐	☐
4. Foreign aid?	☐	☐	☐	☐	☐

I am going to name some institutions in this country. As far as the *PEOPLE RUNNING* these institutions are concerned, would you say that you have a great deal of confidence, only some confidence, or hardly any confidence at all in them? (**CODE FOR EACH ITEM; CODE DK FOR "DON'T KNOW" OR NA FOR "NO ANSWER" IF APPROPRIATE**.)

	Great Deal	Only Some	Hardly Any	DK	NA
5. Education	☐	☐	☐	☐	☐
6. Press	☐	☐	☐	☐	☐
7. Congress	☐	☐	☐	☐	☐

8. Could you tell me, in your own words, how you feel about capital punishment?

That completes the interview. Thank you for your cooperation.

Case ID_____   Respondent's gender: ☐ Male ☐ Female

Hello. My name is _____ and I'm conducting a short survey as part of an exercise for my research methods class. Would you be willing to participate in the study? The interview will take less than five minutes. (**ENCOURAGE IF RELUCTANT.**)

We are faced with many problems in this country, none of which can be solved easily or inexpensively. I'm going to name some of these problems, and for each one I'd like you to tell me whether you think we're spending too much money on it, too little money, or about the right amount. Are we spending too much, too little, or about the right amount on . . . (**READ EACH ITEM AND CHECK THE APPROPRIATE CATEGORY; CODE DK FOR "DON'T KNOW" OR NA FOR "NO ANSWER" IF APPROPRIATE.**)

	Too little	About right	Too much	DK	NA
1. Welfare?	☐	☐	☐	☐	☐
2. Halting the rising crime rate?	☐	☐	☐	☐	☐
3. Improving the nation's education system?	☐	☐	☐	☐	☐
4. Foreign aid?	☐	☐	☐	☐	☐

I am going to name some institutions in this country. As far as the *PEOPLE RUNNING* these institutions are concerned, would you say that you have a great deal of confidence, only some confidence, or hardly any confidence at all in them? (**CODE FOR EACH ITEM; CODE DK FOR "DON'T KNOW" OR NA FOR "NO ANSWER" IF APPROPRIATE.**)

	Great Deal	Only Some	Hardly Any	DK	NA
5. Education	☐	☐	☐	☐	☐
6. Press	☐	☐	☐	☐	☐
7. Congress	☐	☐	☐	☐	☐

8. Could you tell me, in your own words, how you feel about capital punishment?

That completes the interview. Thank you for your cooperation.

Case ID_____   Respondent's gender: ☐ Male ☐ Female

Hello. My name is _____ and I'm conducting a short survey as part of an exercise for my research methods class. Would you be willing to participate in the study? The interview will take less than five minutes. (**ENCOURAGE IF RELUCTANT.**)

We are faced with many problems in this country, none of which can be solved easily or inexpensively. I'm going to name some of these problems, and for each one I'd like you to tell me whether you think we're spending too much money on it, too little money, or about the right amount. Are we spending too much, too little, or about the right amount on . . . (**READ EACH ITEM AND CHECK THE APPROPRIATE CATEGORY; CODE DK FOR "DON'T KNOW" OR NA FOR "NO ANSWER" IF APPROPRIATE.**)

	Too little	About right	Too much	DK	NA
1. Welfare?	☐	☐	☐	☐	☐
2. Halting the rising crime rate?	☐	☐	☐	☐	☐
3. Improving the nation's education system?	☐	☐	☐	☐	☐
4. Foreign aid?	☐	☐	☐	☐	☐

I am going to name some institutions in this country. As far as the *PEOPLE RUNNING* these institutions are concerned, would you say that you have a great deal of confidence, only some confidence, or hardly any confidence at all in them? (**CODE FOR EACH ITEM; CODE DK FOR "DON'T KNOW" OR NA FOR "NO ANSWER" IF APPROPRIATE.**)

	Great Deal	Only Some	Hardly Any	DK	NA
5. Education	☐	☐	☐	☐	☐
6. Press	☐	☐	☐	☐	☐
7. Congress	☐	☐	☐	☐	☐

8. Could you tell me, in your own words, how you feel about capital punishment?

That completes the interview. Thank you for your cooperation.

Case ID_____   Respondent's gender: ☐ Male ☐ Female

Hello. My name is _____ and I'm conducting a short survey as part of an exercise for my research methods class. Would you be willing to participate in the study? The interview will take less than five minutes. (**ENCOURAGE IF RELUCTANT.**)

We are faced with many problems in this country, none of which can be solved easily or inexpensively. I'm going to name some of these problems, and for each one I'd like you to tell me whether you think we're spending too much money on it, too little money, or about the right amount. Are we spending too much, too little, or about the right amount on . . . (**READ EACH ITEM AND CHECK THE APPROPRIATE CATEGORY; CODE DK FOR "DON'T KNOW" OR NA FOR "NO ANSWER" IF APPROPRIATE.**)

	Too little	About right	Too much	DK	NA
1. Welfare?	☐	☐	☐	☐	☐
2. Halting the rising crime rate?	☐	☐	☐	☐	☐
3. Improving the nation's education system?	☐	☐	☐	☐	☐
4. Foreign aid?	☐	☐	☐	☐	☐

I am going to name some institutions in this country. As far as the *PEOPLE RUNNING* these institutions are concerned, would you say that you have a great deal of confidence, only some confidence, or hardly any confidence at all in them? (**CODE FOR EACH ITEM; CODE DK FOR "DON'T KNOW" OR NA FOR "NO ANSWER" IF APPROPRIATE.**)

	Great Deal	Only Some	Hardly Any	DK	NA
5. Education	☐	☐	☐	☐	☐
6. Press	☐	☐	☐	☐	☐
7. Congress	☐	☐	☐	☐	☐

8. Could you tell me, in your own words, how you feel about capital punishment?

That completes the interview. Thank you for your cooperation.

Case ID_____   Respondent's gender: ☐ Male ☐ Female

Hello. My name is _____ and I'm conducting a short survey as part of an exercise for my research methods class. Would you be willing to participate in the study? The interview will take less than five minutes. (**ENCOURAGE IF RELUCTANT**.)

We are faced with many problems in this country, none of which can be solved easily or inexpensively. I'm going to name some of these problems, and for each one I'd like you to tell me whether you think we're spending too much money on it, too little money, or about the right amount. Are we spending too much, too little, or about the right amount on . . . (**READ EACH ITEM AND CHECK THE APPROPRIATE CATEGORY; CODE DK FOR "DON'T KNOW" OR NA FOR "NO ANSWER" IF APPROPRIATE**.)

	Too little	About right	Too much	DK	NA
1. Welfare?	☐	☐	☐	☐	☐
2. Halting the rising crime rate?	☐	☐	☐	☐	☐
3. Improving the nation's education system?	☐	☐	☐	☐	☐
4. Foreign aid?	☐	☐	☐	☐	☐

I am going to name some institutions in this country. As far as the *PEOPLE RUNNING* these institutions are concerned, would you say that you have a great deal of confidence, only some confidence, or hardly any confidence at all in them? (**CODE FOR EACH ITEM; CODE DK FOR "DON'T KNOW" OR NA FOR "NO ANSWER" IF APPROPRIATE**.)

	Great Deal	Only Some	Hardly Any	DK	NA
5. Education	☐	☐	☐	☐	☐
6. Press	☐	☐	☐	☐	☐
7. Congress	☐	☐	☐	☐	☐

8. Could you tell me, in your own words, how you feel about capital punishment?

That completes the interview. Thank you for your cooperation.

Case ID_____   Respondent's gender: ☐ Male ☐ Female

Hello. My name is _____ and I'm conducting a short survey as part of an exercise for my research methods class. Would you be willing to participate in the study? The interview will take less than five minutes. (**ENCOURAGE IF RELUCTANT.**)

We are faced with many problems in this country, none of which can be solved easily or inexpensively. I'm going to name some of these problems, and for each one I'd like you to tell me whether you think we're spending too much money on it, too little money, or about the right amount. Are we spending too much, too little, or about the right amount on . . . (**READ EACH ITEM AND CHECK THE APPROPRIATE CATEGORY; CODE DK FOR "DON'T KNOW" OR NA FOR "NO ANSWER" IF APPROPRIATE.**)

	Too little	About right	Too much	DK	NA
1. Welfare?	☐	☐	☐	☐	☐
2. Halting the rising crime rate?	☐	☐	☐	☐	☐
3. Improving the nation's education system?	☐	☐	☐	☐	☐
4. Foreign aid?	☐	☐	☐	☐	☐

I am going to name some institutions in this country. As far as the *PEOPLE RUNNING* these institutions are concerned, would you say that you have a great deal of confidence, only some confidence, or hardly any confidence at all in them? (**CODE FOR EACH ITEM; CODE DK FOR "DON'T KNOW" OR NA FOR "NO ANSWER" IF APPROPRIATE.**)

	Great Deal	Only Some	Hardly Any	DK	NA
5. Education	☐	☐	☐	☐	☐
6. Press	☐	☐	☐	☐	☐
7. Congress	☐	☐	☐	☐	☐

8. Could you tell me, in your own words, how you feel about capital punishment?

That completes the interview. Thank you for your cooperation.

Case ID_____   Respondent's gender: ☐ Male ☐ Female

Hello. My name is _____ and I'm conducting a short survey as part of an exercise for my research methods class. Would you be willing to participate in the study? The interview will take less than five minutes. (**ENCOURAGE IF RELUCTANT**.)

We are faced with many problems in this country, none of which can be solved easily or inexpensively. I'm going to name some of these problems, and for each one I'd like you to tell me whether you think we're spending too much money on it, too little money, or about the right amount. Are we spending too much, too little, or about the right amount on . . . (**READ EACH ITEM AND CHECK THE APPROPRIATE CATEGORY; CODE DK FOR "DON'T KNOW" OR NA FOR "NO ANSWER" IF APPROPRIATE**.)

	Too little	About right	Too much	DK	NA
1. Welfare?	☐	☐	☐	☐	☐
2. Halting the rising crime rate?	☐	☐	☐	☐	☐
3. Improving the nation's education system?	☐	☐	☐	☐	☐
4. Foreign aid?	☐	☐	☐	☐	☐

I am going to name some institutions in this country. As far as the *PEOPLE RUNNING* these institutions are concerned, would you say that you have a great deal of confidence, only some confidence, or hardly any confidence at all in them? (**CODE FOR EACH ITEM; CODE DK FOR "DON'T KNOW" OR NA FOR "NO ANSWER" IF APPROPRIATE**.)

	Great Deal	Only Some	Hardly Any	DK	NA
5. Education	☐	☐	☐	☐	☐
6. Press	☐	☐	☐	☐	☐
7. Congress	☐	☐	☐	☐	☐

8. Could you tell me, in your own words, how you feel about capital punishment?

That completes the interview. Thank you for your cooperation.

Case ID_____   Respondent's gender: ☐ Male ☐ Female

Hello. My name is _____ and I'm conducting a short survey as part of an exercise for my research methods class. Would you be willing to participate in the study? The interview will take less than five minutes. (**ENCOURAGE IF RELUCTANT.**)

We are faced with many problems in this country, none of which can be solved easily or inexpensively. I'm going to name some of these problems, and for each one I'd like you to tell me whether you think we're spending too much money on it, too little money, or about the right amount. Are we spending too much, too little, or about the right amount on . . . (**READ EACH ITEM AND CHECK THE APPROPRIATE CATEGORY; CODE DK FOR "DON'T KNOW" OR NA FOR "NO ANSWER" IF APPROPRIATE.**)

	Too little	About right	Too much	DK	NA
1. Welfare?	☐	☐	☐	☐	☐
2. Halting the rising crime rate?	☐	☐	☐	☐	☐
3. Improving the nation's education system?	☐	☐	☐	☐	☐
4. Foreign aid?	☐	☐	☐	☐	☐

I am going to name some institutions in this country. As far as the *PEOPLE RUNNING* these institutions are concerned, would you say that you have a great deal of confidence, only some confidence, or hardly any confidence at all in them? (**CODE FOR EACH ITEM; CODE DK FOR "DON'T KNOW" OR NA FOR "NO ANSWER" IF APPROPRIATE.**)

	Great Deal	Only Some	Hardly Any	DK	NA
5. Education	☐	☐	☐	☐	☐
6. Press	☐	☐	☐	☐	☐
7. Congress	☐	☐	☐	☐	☐

8. Could you tell me, in your own words, how you feel about capital punishment?

That completes the interview. Thank you for your cooperation.

EXERCISE 9.4

Name _____

Develop a questionnaire to be given to students on your campus. Write questions on the following concepts:

1. Class level
2. College/division/school (Arts and Science, Business, etc.)
3. Gender
4. Attitudes toward gun control (two questions)
5. Attitudes toward various services provided on your campus (e.g., recreational services, health care, etc.) (three questions)
6. Number of movies seen
7. Rating of movie enjoyment
8. Problems respondents feels are most serious in America today
9. Attitudes toward one or two political issues that you believe may be relevant to students on your campus (two questions)
10. Two questions of your choice

Be sure to include instructions, one contingency question, and some matrix questions. Be sure your questionnaire reflects the advice offered in this chapter.

EXERCISE 9.5

Name _____

Visit the Web site of the American Sociological Association (www.asanet.org). Click "Sociologists." Review the publicly available data sets described under "Data Resources Available" and select one (other than the General Social Survey) that interests you. Or select a data set of interest at the ICPSR Web site (www.icpsr.umich.edu/) besides the General Social Survey.

1.   Describe the data set and where you found it.

2.   Describe a study that you could do using this data set. Identify appropriate independent and dependent variables for your study.

3.   Assess the validity of the data set that you examined.

EXERCISE 9.6

Name _____

Use SPSS or another data analysis program as indicated by your instructor to code the questionnaires you completed in Exercise 9.3. Either develop the code categories according to the instructions provided in the chapter or follow your instructor's advice.

Describe any problems you encountered and how you resolved them.

ADDITIONAL INTERNET EXERCISES

1. Visit the American Statistical Association's website at http://www.whatisasurvey.info/, click "Click here to read Chapter 1," and then click on "Chapter 6." Read this chapter, "Designing a Questionnaire." Summarize the issues in that segment and determine how the ASA's advice on questionnaire construction meshes with Babbie's advice.

2. Access the site http://www.busreslab.com/tips/tipsgen.htm, which is produced by the Business Research Lab. This site gives "tips" on such things as leading questions and minimizing non-response. Summarize three of the topics and show how this company's statements compare with the text on how to handle the selected issues.

3. Visit http://www.busreslab.com/onlineempsurveydemo.htm, which is also from the Business Research Lab, and demonstrates an online survey on employee satisfaction. Critique each question using the guidelines suggested in the text.

4. The Census is perhaps the most well known survey. Visit its Web site at http://www.census.gov/ and click on "American Fact Finder." Input your home town zip code and view the results. What is the most

Chapter 10

Qualitative Field Research

OBJECTIVES

1. Define qualitative field research and compare it with other methods.

2. Identify the key strengths of field research.

3. Define and give examples of each of the following elements of social life appropriate for field research: practices, episodes, encounters, roles and social types, social and personal relationships, groups and cliques, organizations, settlements and habitats, and subcultures and lifestyles.

4. Give three examples of research topics particularly appropriate for field research.

5. Compare the various roles the field researcher can assume, ranging from complete participant to complete observer.

6. Note the advantages and disadvantages of adopting the views of the people studied while doing field research.

7.   Compare the postures of fully accepting the beliefs, attitudes, and behaviors of those under study versus remaining more "objective."

8.   Describe and note the advantages of each of the following field research paradigms: naturalism, ethnomethodology, grounded theory, case studies, the extended case method, institutional ethnography, and participatory action research.

9.   Provide advice on each of the following steps in preparing for the field: review of relevant literature, use of informants, and establishing initial contacts.

10.   Provide advice for asking questions in field research, and compare a field research interview with normal conversation.

11.   Describe the stages in a complete interviewing process: thematizing, designing, interviewing, transcribing, analyzing, verifying, and reporting.

12.   Show how focus groups are relevant in field research.

13.   Identify relevant ethical concerns in qualitative field research.

14.   Discuss the strengths and weaknesses of field research, particularly in terms of validity and reliability.

OUTLINE

1.   Topics appropriate to field research

2.   Special considerations in qualitative field research
   a.   The various roles of the observer
   b.   Relations to subjects

3.   Some qualitative field research paradigms
   a.   Naturalism
   b.   Ethnomethodology
   c.   Grounded theory

SUMMARY

Field research closely parallels our daily observation of and participation in social behavior, as well as our attempt to understand such behavior. The social scientific application of field research requires specific skills and techniques and hence is more useful than casual observation. It is both a data-gathering strategy and a theory-generating activity. The term "qualitative field research" is used to distinguish this type of observation method from more quantitative methods (such as surveys).

Field research is particularly appropriate for topics that appear to defy simple quantification and that require the comprehensiveness of perspective the field researcher can provide. It is also useful for studying topics and people within their natural settings and for investigating social processes over time.

The field researcher may investigate various types of social phenomena. One type is practices, which involve behavior. Episodes refer to specific events. Encounters involve people interacting. Roles and social types include positions and behavior. Another type is social and personal relationships, which pertain to behavior found in sets of roles. Groups and cliques involve small numbers of interacting people. Organizations are larger than groups. Settlements and habitats include small-scale "societies." Finally, subcultures and lifestyles are examined when field researchers focus on how large numbers of people adjust to life.

The field researcher's role may vary according to the degree of participation. The researcher's identity and purpose are not known to those being observed in the complete participant role. This strategy requires complete adherence to the norms of the group but may involve ethical problems regarding the deception involved in this approach. Also, people being observed might modify their behavior if they know they are being studied, a problem known as reactivity. The researcher playing the complete observer role observes a social situation without becoming a part of it.

Field researchers are often caught in a dilemma regarding their relations to their subjects. On the one hand, they may choose to actually take on the beliefs, attitudes, and behaviors of those being studied. Doing so enhances insider understanding but comes at the cost of losing the opportunity to see and understand issues within frames of reference unavailable to the subjects. On the other hand, field researchers may choose to remain more distant in their struggle to remain more objective. Assuming both postures yields important advantages.

Several field research paradigms exist. These paradigms do not differ on specific methods used, but they do differ on what the data mean (epistemological differences). Naturalism is an older approach and is based on the positivist notion that social reality is "out there," ready to be naturally observed and reported by the researcher as it "really is." The naturalist approach strives to tell people's stories the way they "really are" instead of the way the researchers understands "them." An ethnography is a study that focuses on detailed and accurate description rather than explanation, and is a form of naturalism.

Ethnomethodology is based on the phenomenological premise that reality is socially constructed rather than being "out there" for us to observe. That is, people describe their experiences not "as they are" but "as they make sense of them." Hence, researchers cannot rely on their subjects' stories to depict social realities accurately. Ethnomethodologists study the taken-for-granted expectations that people have about everyday life. One way to study these expectations is to violate the social rules in order to highlight the subtleties of everyday interaction.

Grounded theory is the attempt to derive theories from an analysis of the patterns, themes, and common categories discovered in observational data. It combines a naturalistic approach with a positivist concern for systematic procedures. It allows the researcher to be scientific and creative at the same time. Some guidelines include: think comparatively, obtain multiple viewpoints, periodically step back, maintain an attitude of skepticism, and follow the research procedures.

Case studies focus attention on one or a few instances of some social phenomenon. The term "case" is used broadly. Case studies can be both descriptive and explanatory. The goal can be an idiographic understanding of a particular case or they can help develop more general, nomothetic theories. The extended case method seeks to discover flaws in, and then modifying, existing social theories.

An institutional ethnography examines the ideologies that shape the experiences of oppressed subjects. If researchers ask those in subordinated groups about "how things work," they can discover the institutional practices that shape their realities. This approach helps uncover forms of oppression that often are overlooked by more traditional types of research. Finally, the researcher doing participatory action research serves as a resource to those being studied, typically disadvantaged groups, so that those being studied can act in their own interest. In such studies, those studied play a major role in designing the research to help address their goals. But this approach carries a risk that it will be unclear who is in charge: the researcher or the empowered participants. Emancipatory research involves producing knowledge with political outcomes by benefitting oppressed people.

Field research follows a logical progression of steps. The researcher must first prepare for the field by examining relevant literature, by using informants to help frame the analysis, and by establishing initial contacts with the people to be studied. Field researchers typically employ qualitative interviews to ask questions. This conversational approach yields flexibility in that an answer to one question may influence the next question the researcher wishes to ask. Effective probing is essential. It also helps to assume the role of the "socially acceptable incompetent."

Qualitative interviewing process involves seven stages. Thematizing helps clarify the purpose of the interviews and concepts to be studied. Designing involves laying out the processes of the research. Interviewing involves doing the actual interviews. Transcribing creates a written text. Analyzing determines the meaning of materials gathered relative to the study. Verifying involves checking the reliability and validity of the materials. Finally, through reporting the researcher shares what was learned.

With focus groups, researchers bring people into their study headquarters for qualitative interviewing and observation. A type of group interviewing, the focus group approach is based on structured, semi-structured, or unstructured interviews. It is typically used in marketing research. Typically 12 to 15 people are brought together to engage in a guided discussion of a topic. This method is flexible, has high face validity, produces speedy results, and is inexpensive. But this method also yields less control than when only one person is interviewed, sometimes provides data that are difficult to analyze, requires special skills, sometimes yields problems among groups, involves difficulties in assembling groups, and requires a conducive environment.

The field journal contains the field researcher's observations. Field notes should include both empirical observations as well as interpretations, although the observations and interpretations should be kept separate. Field researchers learn not to trust their memories and hence take many notes at multiple stages in the research process. They record as much as possible so that interpretations that do not seem evident at the time the notes are taken may emerge later.

Field research does raise ethical concerns. Particularly relevant are issues surrounding permission of those studied, determining appropriate levels of involvement, deciding what positions to take, and the role of informants.

The greatest strength of the field research method lies in the presence of an observing, thinking researcher on the scene of the action. Field researchers are in a unique position to examine the nuances of attitudes and behaviors and to examine social processes over time; both contribute to a deeper understanding. Field research also is inexpensive and provides a great deal of flexibility for moving between observation and analysis. However, field research seldom yields descriptive statements about large populations because it is a qualitative rather than a quantitative approach. As a result, the field researcher's conclusions are seen as more tentative rather than definitive. Although field research is often more valid than surveys and experiments, it does suffer from lower reliability.

TERMS

1. case studies	9. naturalism
2. emancipatory research	10. qualitative field research
3. ethnography	11. qualitative interview
4. extended case method	12. participatory action research
5. ethnomethodology	13. reactivity
6. focus groups	14. reflexivity
7. grounded theory	15. symbolic realism
8. institutional ethnography	

MATCHING

_____ 1. An approach that examines the rules that govern everyday life, often by breaking the rules.

_____ 2. An approach based on the assumption that social reality is "out there," ready to be observed and reported by the researcher as it "really is."

_____ 3. Deriving theories from an analysis of the patterns, themes, and common categories discov-ered in observational data.

_____ 4. An approach that focuses attention on one or a few instances of some social phenomenon.

_____ 5. The purpose of this approach is to discover flaws in, and to modify, existing social theories.

_____ 6. An approach in which members of subordinated groups are asked about "how things work" so that researchers can discover the institutional practices that shape their realities.

_____ 7. With this paradigm, the researcher's function is to serve as a resource to those being studied, typically disadvantaged groups, as an opportunity for them to act effectively in their own interest.

_____ 8. A method in which about 12 to 15 people are brought together to engage in a guided discussion of some topic.

_____ 9. A method for producing knowledge with political outcomes by benefiting oppressed people.

_____ 10. The term given to situations where subjects might alter their behavior if they know that they are being studied.

TRUE-FALSE QUESTIONS

T F 1. The complete participant role (especially when the researcher's identity is not known) is more likely than the complete observer role to avoid reactivity problems.

T F 2. The complete observer role is more likely than the complete participant role to yield ethical problems.

T F 3. Symbolic realism reflects the need for researchers to treat the beliefs they study as worthy of respect rather than as objects of ridicule.

T F 4. Naturalism is the paradigm that emphasizes breaking the rules so that people's taken-for-granted expectations would become apparent.

T F 5. Grounded theory is a method of observational research that emphasizes that researchers remain grounded within their disciplinary perspectives.

T F 6. The extended case method is particularly useful for discovering flaws in, and then modifying, existing social theories.

T F 7. Participatory action research involves bringing about 12 to 15 people together to engage in a guided discussion of some topic.

T F 8. Field research seems to provide measures with greater validity than do survey and experimental measurements.

T   F   9.   Emancipatory research is a design that frees the researcher from some of the usual ethical obligations in doing field research.

T   F   10.   One of the key strengths of field research is how comprehensive a perspective it can give researchers.

REVIEW QUESTIONS

1. Field research differs from other modes of observation in that
   a. it is more highly refined.
   b. it is both a data-collecting activity and a theory-generating activity.
   c. it involves more careful conceptualization and operationalization.
   d. it is more dependent on using deduction to generate specific hypotheses for testing.
   e. it is more reliable and valid.

2. Professor Sequoia did a study in which she performed a detailed and accurate description, rather than an explanation, of the social interactions at a church. Which approach did she use?
   a. ethnomethodology
   b. ethnography
   c. participatory action research
   d. focus group
   e. case study

3. Professor Oxley wanted to examine how new teachers make sense of their everyday worlds. He did so by asking some of them to break some rules so that he could better uncover the norms. Which approach did he use?
   a. grounded theory
   b. extended case method
   c. focus group
   d. participatory action research
   e. ethnomethodology

4.   Sonja wanted to do a study of playground interaction among children but, instead of beginning with a theory and developing hypotheses, she wanted to spend a few months observing and then let the patterns or common categories emerge. Which approach did she use?
  a.   ethnography
  b.   naturalism
  c.   qualitative interviews
  d.   grounded theory
  e.   focus group

5.   Professor Sullivan performed an observational study of the norms that govern interactions between cab drivers and their passengers. Which one of the following does this example reflect?
  a.   roles
  b.   encounters
  c.   episodes
  d.   groups
  e.   settlements

6.   The most serious ethical problems exist for which type of research role?
  a.   complete participant
  b.   participant-as-observer
  c.   observer-as-participant
  d.   complete observer
  e.   ethnographer

7.   A common first step in doing field research is
  a.   considering ethics.
  b.   forming a hypothesis.
  c.   constructing a theory.
  d.   deciding on level of participation.
  e.   reviewing literature.

8. Which one of the following is *most* appropriate for asking questions during field research?
   a. structured questionnaire
   b. unstructured questionnaire
   c. structured interviews
   d. qualitative interviews
   e. projective techniques

9. Which one of the following is *false* regarding field notes? Or are they all true?
   a. Don't trust your memory more than you have to.
   b. Take notes in stages.
   c. Get the major points, but don't worry about getting as many details as you can.
   d. Rewrite your notes before going to sleep.
   e. All are true.

10. Professor Patecki studied a ghetto in a big city as part of his field research. Which element did he study?
    a. practices
    b. lifestyles
    c. settlement
    d. relationships
    e. episodes

11. Professor Johnson wants to ask members of a subordinated group about how things work in their lives so that he can identify the structural practices that shape their realities. Which approach is best?
    a. institutional ethnography
    b. focus group.
    c. extended case method
    d. participatory action research
    e. ethnography

12. Francesca wanted to study homeless women, but in a way that allowed the homeless women to define the research problem, define the remedies desired, and take the lead in designing the research that will help them realize their aims. Which approach should she use?
    a. participatory action research
    b. case study
    c. extended case method
    d. grounded theory
    e. naturalism

13. In comparison to surveys and experiments, field research has
    a. high validity and high reliability.
    b. high validity and low reliability.
    c. low validity and low reliability.
    d. low validity and high reliability.
    e. high reliability, but only when the validity is high.

14. Jody wants to do a study of depressed welfare mothers but wishes to change the political issues leading to reduced welfare benefits. She wants her research to benefit oppressed welfare recipients. Best to use would be
    a. naturalism.
    b. grounded theory.
    c. emancipatory research.
    d. institutional ethnography.
    e. focus groups.

15. Ernestine desires to study how the social interactions of residents in a nursing home change over a five-year period. Which benefit of field research best applies to her case?
    a. it is comprehensive
    b. it occurs in natural settings
    c. it is relatively free of ethical concerns
    d. it is cheap
    e. it is well suited to the study of social processes over time

16.   Which of the following is a study that focuses on detailed and accurate description rather than explanation?
  a.   ethnography
  b.   ethnomethodology
  c.   participatory action research
  d.   emancipatory research
  e.   case studies

17.   The goal in extended case method research is to
  a.   provide an idiographic understanding of a particular case.
  b.   discover flaws in, and then modifying, existing social theories.
  c.   derive theories from an analysis of patterns and categories discovered in observational data.
  d.   reduce reactivity.
  e.   reduce the negative ethical impacts of field research.

18.   Professor Gabino did a field research study of softball teams. Which element of social life did he examine?
  a.   practices
  b.   episodes
  c.   encounters
  d.   relationships
  e.   groups

19.   Professor Rosenblatt did a field research study of urban underclass in Chicago, those inner city residents left without jobs when industries moved overseas or to the suburbs. Which element of social life did he examine?
  a.   practices
  b.   organizations
  c.   encounters
  d.   lifestyles/subcultures
  e.   groups

20.   Which stage in the interviewing process involves clarifying the purpose of the interviews and the concepts to be explored?
   a.   thematizing
   b.   designing
   c.   interviewing
   d.   transcribing
   e.   analyzing

DISCUSSION QUESTIONS

1. Explain what Babbie means when he says that field research is not only a data-gathering activity but a theory-generating activity as well. Illustrate each function. Which is more important? Why?

2. Compare and contrast the types of qualitative inquiry: naturalism, ethnomethodology, grounded theory, case studies, the extended case method, institutional ethnography, and participatory action research. Select a topic and show how this topic would be studied with each approach.

3. Babbie discusses the following stages in field research: preparation for the field, qualitative interviewing, focus groups, and recording observations. Select three and discuss the basic principles involved for effectively completing each.

4. Compare the strengths and weaknesses of field research with both surveys and experiments, particularly in terms of reliability and validity.

EXERCISE 10.1

Name _____

Your assignment is to observe and understand the social dynamics of jaywalking—defined narrowly for our purposes as "pedestrians crossing the street against the light at intersections."

You are to make three separate observations of at least 15 minutes duration each.

Your specific tasks are to discover:

1. What are the characteristics of people who are the most likely to jaywalk and what are the characteristics of those least likely to jaywalk?

2. What are the characteristics of people who are likely to stimulate jaywalking by others? That is, what are the characteristics of those, when they jaywalk, who seem to be followed by people who were previously waiting for the light to change?

Record your field notes below.

Please don't get arrested for jaywalking!

FIELD NOTES:

EXERCISE 10.2

Name _____

Your assignment is to pick some location or situation in which strangers are put in physical proximity to one another for a period of time. Examples include a bus stop, a coffee shop, a table at the library, a train station or airport waiting room, a hospital waiting room, a party, etc.

Observe at your location for at least an hour. Examine the ways in which strangers interact with one another. Notice the ways in which they communicate: through words, facial expressions, body postures, and the like. Examine what is communicated, in what situations, with what effect, and by what kinds of people. Examine also the importance of social space and physical space. Pay special attention to (a) the varieties in all these things, and (b) any regular patterns that seem to exist.

Record your field notes below.

FIELD NOTES:

EXERCISE 10.3

Name _____

Your assignment is to plan a possible field research project to investigate differences in friendship patterns in coed and single-sex residence halls. The questions below deal with some of the concepts you may wish to consider in such a study. For each, describe some of the possible indicators that might be relevant to the concept. You can observe and you can do unstructured interviews using a field notebook, but you cannot hand out a questionnaire or do formal interviews using structured interview schedules. You want to be able to determine if the friendships are with people in the same dorm.

1. How would you determine the number of friendships each resident in the coed and single-sex residence halls has (friendships in the same hall in which the resident lives)?

2. How would you determine the quality of friendships? Think of quality as the "strength" or "intensity" of friendship. How would you determine if the friendships are close (strong), distant (weak), or in between?

(continued)

3. Which other aspects of friendship patterns would you expect to differ in the two situations? List at least three along with possible indicators. Think about such activities as dating, recreation, visiting, topics of conversation, and the like, but be sure to be specific and describe how you would expect them to differ in the two types of residence halls (coed or single-sex) residence halls. What would be some indicators that you could observe to determine if they do differ?

4. List two other variables that you might consider in a study such as this. These should be variables besides residence hall type (coed or not) that might be related to the number and quality of friendships that develop. Think about what might be some variables that would determine whom residents become friends with or determine how strong a friendship would develop. Explain why.

(continued)

5. Describe how you might measure the first variable you noted in #4. What would you be looking for, or what types of questions might you ask?

6. Describe how you might measure the second variable in #4. What would you be looking for or what types of questions might you ask?

EXERCISE 10.4
Name _____

Review either or both of the following Web sites on qualitative analysis and find a field research study:

> www.qualitative-research.net/index.php/fqs/index
> www.qualitativeresearch.uga.edu/QualPage/

1.   Citation:

2.   Summarize the study.

3.   Which of the following are represented in the study: naturalism, ethnomethodology, grounded theory, case studies, the extended case method, institutional ethnography, and participatory action research? Explain how.

4.   Which of the following elements of social life are examined in the study: practices, episodes, encounters, roles, relationships, groups, organizations, settlements, social worlds, and lifestyles (or subcultures)? Explain.

(continued)

5.   How were the data analyzed? Present the major conclusions.

6.   Address any ethical issues that emerged.

7.   Explain why the field research method was the most appropriate method to use for this study.

ADDITIONAL INTERNET EXERCISES

1. Access the University of Michigan's Center for Ethnography of Everyday Life web site at http://ceel.psc.isr.umich.edu/ and click on "Projects" on the left. Select a project that is currently underway. Summarize the description of the project and explain why the study is an ethnography.

2. Access the website for the Qualitative Research Consultants Association at http://www.qrca.org/ and make at least three connections between the content at this site and the information presented in the Babbie text.

3. Scan the qualitative journals listed at http://www.slu.edu/organizations/qrc/QRjournals.html and select two studies to summarize. Also show how each study connects with content in the Babbie chapter.

Chapter 11

Unobtrusive Research

OBJECTIVES

1. Describe and compare the three unobtrusive research designs: content analysis, analysis of existing statistics, and historical/comparative analysis.

2. Give three examples of artifacts that content analysts might study.

3. Give three examples of content analysis in which the unit of observation differs from the unit of analysis.

4. Show how the unit of analysis influences sample selection in content analysis.

5. Illustrate how a researcher might employ each of the following sampling techniques in content analysis: simple random sampling, systematic sampling, stratified sampling, and cluster sampling.

6. Differentiate manifest content from latent content by definition and example.

7.   Present advice for the development of code categories in content analysis.

8.   Present advice for counting and record keeping in content analysis.

9.   Outline the strengths and weaknesses of content analysis.

10. Explain how analytic induction is used in qualitative content analysis.

11. Summarize the difficulties with units of analysis in existing statistics.

12. Explain why validity is a problem with existing statistics, and present two strategies for resolving this problem.

13. Explain why reliability is a problem with existing statistics, and present two strategies for resolving this problem.

14. List three sources of existing statistics.

15. List three sources of data for historical/comparative analysis.

16. Discuss the role of corroboration in enhancing the quality of existing statistics.

17. Discuss the role of *verstehen*, and ideal types in the analysis of existing statistics.

18. Indicate the ethical issues in using unobtrusive measures

OUTLINE

1.   Content analysis
   a.   Topics appropriate to content analysis
   b.   Sampling in content analysis
   c.   Coding in content analysis
   d.   An illustration of content analysis
   e.   Strengths and weaknesses of content analysis

2.   Analyzing existing statistics
       a.   Durkheim's study of suicide
       b.   Consequences of globalization
       c.   Units of analysis
       d.   Problems of validity
       e.   Problems of reliability
       f.   Sources of existing statistics

3.   Comparative and historical research
       a.   Examples of comparative and historical research
       b.   Sources of comparative and historical data
       c.   Analytical techniques

4.   Ethics and unobtrusive measures

SUMMARY

Unlike other methods, unobtrusive methods do not generally influence the data gathered. Content analysis is one type of unobtrusive method. It may be applied to virtually any form of communication and involves examining artifacts to answer the classic question: "Who says what, to whom, how, and with what effect?"

The complexity of units of analysis is very apparent in content analysis, particularly when the unit of observation differs from the unit of analysis. Content analysts must address this issue directly because sample selection depends largely on the unit of analysis. Any of the conventional sampling techniques described in Chapter 8 may be employed in content analysis, including random or systematic sampling, stratified sampling, and cluster sampling. Sampling need not end when the researcher reaches the unit of analysis.

Content analysts distinguish manifest from latent content. Manifest content is the visible surface content, and its coding approximates the use of a standardized questionnaire. Latent content reflects the underlying meaning, but this advantage comes at a cost of reliability and specificity. It is best to code both types.

Conceptualization plays a key role in determining code categories and frequently requires both inductive and deductive methods. Attributes of variables must be clearly specified, and each of the several levels of measurement may be employed. The coding operations must be amenable to data processing. Hence, the final product must be numerical, units of analysis and units of observation must be clearly distinguished, and the base from which counting is done must be carefully considered and must be presented. Occasionally a qualitative analysis of the materials is more appropriate. Analytic induction can be used for this purpose; it begins with observations to develop hypotheses but goes beyond description to uncover patterns and relationships among variables. The main danger in this procedure is that the researcher may misclassify observations to support emerging hypotheses.

The greatest advantage of content analysis is economy in terms of time and energy. Content analysis is also quite safe—mistakes can be easily rectified and the study redone. Content analysis also allows the study of processes over time. Finally, content analysis is unobtrusive. But it is limited to an examination of recorded communications. Content analysis has both advantages and disadvantages in terms of validity and reliability.

Many governmental and other agencies collect official or quasi-official statistics. Existing statistics can provide the main data for a social scientific study and can also provide supplemental sources of data for studies employing other research designs. The unit of analysis is frequently not the individual but instead is found in different types of aggregates. The aggregate nature of existing statistics sometimes presents a problem, requiring the social scientist to make inferences from one level to another. Doing so, however, may result in committing the ecological fallacy—generalizing from data gathered at one unit of analysis to another unit of analysis.

Existing statistics are particularly prone to problems of validity because the data may not reflect a particular measure that a social scientist has constructed. This problem may be addressed through logical reasoning and replication. Reliability may also be a problem because different agencies use different data-gathering strategies with varying levels of accuracy. This problem may be addressed through awareness and logical reasoning, as well as replication. There are many sources of existing statistics, such as *The Statistical Abstract of the United States*, *The Demographic Yearbook*, and Web sites for various governmental agencies.

Historical/comparative research helps trace the development of social forms over time and helps compare such developments across cultures. In its purest form, this approach relies on "primary sources" of data, such as diaries and official government documents. Once again, problems of reliability and bias enter, which again can be addressed through replication (known as corroboration in historical research).

Analysts in this tradition often stress the role of *verstehen*, an understanding of the essential quality of a social phenomenon. Sometimes ideal types help researchers find patterns in the details of their data; ideal types are conceptual models composed of the essential characteristics of social. Historical/comparative research is often performed in light of a particular theoretical paradigm, such as Marxism. Time-series data can be used to do a quantitative analysis of changing conditions over time.

Using unobtrusive measures avoids many of the ethical principles discussed in previous chapters. However, some still apply. Some projects may warrant confidentiality considerations. All study procedures and results must be described honestly.

TERMS

1. analysis of existing statistics)
2. analytic induction
3. coding
4. comparative/historical research
5. content analysis
6. corroboration
7. ecological fallacy
8. ideal types
9. latent content
10. manifest content
11. negative case testingt
12. unobtrusive research
13. *verstehen*

MATCHING

_____ 1. The study of recorded human communications.

_____ 2. The visible, surface content of a communication.

_____ 3. The underlying meaning of a communication.

_____ 4. This occurs when a researcher gathers data from groups and generalizes to individuals.

_____ 5. In historical research the process by which several sources point to the same set of "facts."

_____ 6. Research that allow the researcher to study social life without influencing it in the process.

_____ 7. Understanding in which the researcher mentally takes on the circumstances, views, and feelings of those being studied so that the researcher can interpret their actions appropriately.

_____ 8. Conceptual models composed of the essential characteristics of social phenomena.

_____ 9. The process of transforming raw data into a standardized form.

_____ 10. The examination of societies or other social units over time and in comparison with one another.

TRUE-FALSE QUESTIONS

T   F   1. Content analysis is particularly well suited to studying communications.

T   F   2. Determining the appropriate unit of analysis in content analysis is generally simpler than doing so for research on individuals.

T   F   3. Sampling in content analysis generally occurs only at one level.

T   F   4. Counting the number of times that the masculine pronoun is used in an article is an example of coding manifest content.

T   F   5.   It is important in content analysis record keeping to clearly distinguish between units of analysis and units of observation.

T   F   6.   Durkheim's study of suicide is an example of analyzing existing statistics.

T   F   7.   The United Nation's *Demographic Yearbook* is the single most valuable source of data.

T   F   8.   *Verstehen* reflects the need for comparative/historical researchers to remove themselves from the circumstances and feelings of those being studied to increase the objectivity of the study.

T   F   9.   Comparative and historical research is a quantitative method.

T   F   10. The use of unobtrusive measures avoids many of the ethical issues relevant for other research methods.

REVIEW QUESTIONS

1.   Content analysis may be applied to
   a.   anything that is written.
   b.   any form of communication.
   c.   anything social scientific that does not involve surveys, experiments, or observation.
   d.   anything that has both manifest and latent content.
   e.   anything that involves statistics.

2.   Sampling and data analysis are generally more complex in content analysis because
   a.   the method is less highly developed than other modes of observation.
   b.   both sampling and data analysis are not as appropriate.
   c.   sampling error cannot be calculated.
   d.   current statistical techniques do not apply.
   e.   the units of analysis are not always clear.

3.   Which of the following sampling techniques is/are applicable in content analysis?
   a.   simple random
   b.   systematic
   c.   stratified
   d.   purposive
   e.   all of the above

4.   Professor Hague codes the number of times the word sex is used in commercials. She is examining
   a.   latent structure.
   b.   ecological fallacy.
   c.   manifest content.
   d.   corroboration.
   e.   latent content.

5.   Professor Spykes codes how frequently situation comedies portray sexual activities. He is examining
   a.   latent structure.
   b.   sociometric choices.
   c.   manifest content.
   d.   corroboration.
   e.   latent content.

6.   Which one of the following is *not* a strength of content analysis? Or are they all strengths?
   a.   economy in time and money
   b.   flexibility and safety
   c.   allows study of process over time
   d.   unobtrusive
   e.   all are strengths

7. In studying suicide, Durkheim used
    a. surveys.
    b. experiments.
    c. observation.
    d. government statistics.
    e. evaluation research.

8. Durkheim concluded that suicides are a product of
    a. social instability and disintegration.
    b. temperature and population density.
    c. political party structure.
    d. education and family structure.
    e. depression.

9. The most common unit of analysis involved in the analysis of existing statistics is
    a. individuals.
    b. groups.
    c. organizations.
    d. artifacts.
    e. cities.

10. Because most existing statistics are aggregated, researchers must avoid the error of
    a. overgeneralizing.
    b. validity.
    c. drawing conclusions about individuals.
    d. drawing conclusions about the aggregate.
    e. reductionism.

11. A *major* disadvantage of analyzing existing data is
    a. ethical violations.
    b. errors in coding.
    c. inappropriate statistical techniques.
    d. inappropriate units of analysis.
    e. validity of measures.

12.   The two features of science used to handle the problem of
      validity in analysis of existing statistics are
      a.   cumulativeness and generalizability.
      b.   generalizability and objectivity.
      c.   objectivity and replication.
      d.   replication and logical reasoning.
      e.   logical reasoning and cumulativeness.

13.   One of the most valuable sources of existing statistics is
      a.   *Statistical Abstract of the United States.*
      b.   Superintendent of Documents.
      c.   *TV Guide.*
      d.   *World Book Encyclopedia.*
      e.   Gallup polls.

14.   Professor Dealy performed his historical analysis of race
      relations over the last 100 years and is now at work developing a
      conceptual model composed of the essential characteristics of
      racism. He detailed the core features of racism in his attempt to
      create a theoretical model of "perfect" racism. He used which
      one of the following?
      a.   hermeneutics
      b.   ideal type formation
      c.   *verstehen*
      d.   sociological historical analysis
      e.   paradigm development

15.   One ethical issue that may enter into a study of diaries of
      deceased people is
      a.   voluntary participation.
      b.   confidentiality.
      c.   debriefing.
      d.   no harm to subjects during the experiment.
      e.   convincing an Institutional Review Board that the study
            should occur.

16.   Tanika did a historical study of the speeches given by death row inmates just before their executions. She wanted to take on mentally the circumstances, views, and feelings of the inmates. Which technique did she use?
    a.   hermeneutics
    b.   ideal type formation
    c.   *verstehen*
    d.   sociological historical analysis
    e.   paradigm development

17.   The general goal in analyzing data in comparative and historical studies is to
    a.   look for patterns.
    b.   look at the impact of specific historical events.
    c.   determine the unit of analysis.
    d.   use focus groups.
    e.   use statistics.

18.   The best solution to the dilemma of choosing to analyze either manifest or latent content when doing a content analysis study is to
    a.   choose the one you like the best.
    b.   choose manifest because it is more empirical.
    c.   choose latent because it gets more at the meanings.
    d.   take turns using each one in different studies.
    e.   use both.

19.   Content analysis is essentially what type of operation?
    a.   sampling
    b.   coding
    c.   theory construction
    d.   difficult
    e.   *verstehen*

20. Hickory used content analysis to study letters to the editor in order to determine which cities were more liberal and which were more conservative. What is the unit of observation and what is the unit of analysis?
    a.   cities/letters
    b.   people/cities
    c.   letters/cities
    d.   newspapers/people
    e.   newspapers/cities

DISCUSSION QUESTIONS

1. Compare and contrast content analysis, analysis of existing statistics, and comparative/historical analysis in terms of purposes, strengths, and weaknesses.

2. Explain why units of analysis are of particular concern in content analysis.

3. Explain why social science researchers are interested in both manifest and latent content. Show how these two types of content are both similar and different.

4. Discuss the special problems regarding validity and reliability experienced by those who analyze existing statistics.

EXERCISE 11.1

Name _____

You are to compare three national news magazines in their coverage of some major news story in the past year.

1. First, select a news story that involved different points of view. Note the story here.

2. Next, select three news magazines, from among such publications as *Newsweek, Time, U.S. News & World Report, The New Republic, The Nation, National Review*, etc. (Your instructor may suggest others.) Pick two different issues of each magazine. Indicate the magazines and the dates of the issues you studied.

> Magazine 1
> > Date of first issue studied:
> > Date of second issue studied:

> Magazine 2
> > Date of first issue studied:
> > Date of second issue studied:

> Magazine 3
> > Date of first issue studied:
> > Date of second issue studied:

3. Describe the method you used to determine the relative amount of coverage each magazine devoted to the story you are studying.

(continued)

4. Report your findings regarding the amount of coverage each magazine devoted to the issue. Provide both data tabulations and your interpretations of those data.

Magazine # 1:

Magazine # 2:

Magazine # 3:

5. Describe the method you used to determine the editorial positions taken by the magazines.

6. Report your findings regarding the editorial positions taken by the three magazines. Provide both data tabulations and your interpretations of those data.

Magazine # 1:

Magazine # 2:

Magazine # 3:

EXERCISE 11.2

Name _____

This exercise involves examining television commercials. Select three commercials for analysis. If feasible, record them so you can review them again. Try to pick commercials that have appeared repeatedly. Examine them for their manifest and latent content.

For additional information on video content analysis, see www.lienhart.de (click "Video Content Analysis Homepage").

1. Describe the commercials before beginning your analysis.

   Commercial # 1:

   Commercial # 2:

   Commercial # 3:

(continued)

2. Describe how you assessed the manifest content of each commercial and summarize the manifest content for each.

Commercial # 1:

Commercial # 2:

Commercial # 3:

3. Describe how you assessed the latent content of each commercial and summarize the latent content for each.

Commercial # 1:

Commercial # 2:

Commercial # 3:

EXERCISE 11.3

Name _____

Test the following hypothesis: States with large populations have higher auto theft rates than do states with small populations.

To test this hypothesis, you must determine the population of each of the 50 states in a recent year and the number of auto thefts reported in each state that year. You might check the *Statistical Abstract of the United States* (http://www.census.gov/compendia/statab/).

Complete the worksheet below, and rank order the 50 states in terms of population. Combine them into five groups of ten each, ranging from the ten smallest to the ten largest. Once you have them in groups of ten, combine the population for each group, combine the total number of thefts for each group, and divide the combined number of thefts by the combined population to get the auto theft rate for each group (see step 3).

1.   Give the bibliographical citation of the source(s) from which your data are taken.

2.   Complete the following worksheet:

	POPU-LATION	AUTO THEFTS		POPU-LATION	AUTO THEFTS
Alabama	_____	_____	Connecticut	_____	_____
Alaska	_____	_____	Delaware	_____	_____
Arizona	_____	_____	Florida	_____	_____
Arkansas	_____	_____	Georgia	_____	_____
California	_____	_____	Hawaii	_____	_____
Colorado	_____	_____	Idaho	_____	_____
Illinois	_____	_____	New York	_____	_____

(continued)

Indiana	_____	_____	North Carolina	_____	_____
Iowa	_____	_____	North Dakota	_____	_____
Kansas	_____	_____	Ohio	_____	_____
Kentucky	_____	_____	Oklahoma	_____	_____
Louisiana	_____	_____	Oregon	_____	_____
Maine	_____	_____	Pennsylvania	_____	_____
Maryland	_____	_____	Rhode Island	_____	_____
Massachusetts	_____	_____	South Carolina	_____	_____
Michigan	_____	_____	South Dakota	_____	_____
Minnesota	_____	_____	Tennessee	_____	_____
Mississippi	_____	_____	Texas	_____	_____
Missouri	_____	_____	Utah	_____	_____
Montana	_____	_____	Vermont	_____	_____
Nebraska	_____	_____	Virginia	_____	_____
Nevada	_____	_____	Washington	_____	_____

4. Complete the following table (warning: be sure to convert states'
    auto theft *rates* into *numbers* of auto thefts per state before
    combining state figures):

5.

	COMBINED POPULATION	COMBINED NUMBER OF AUTO THEFTS	COMBINED AUTO THEFT RATE
Smallest ten states	_____	_____	_____
Second smallest ten states	_____	_____	_____
Third smallest ten states	_____	_____	_____
Fourth smallest ten states	_____	_____	_____
Largest ten states	_____	_____	_____

4.   Give a brief research report, indicating whether you feel the hypothesis was confirmed or disconfirmed.

EXERCISE 11.4

Name _____

Select an article reporting on content analysis research (http://www.aber.ac.uk/media/Sections/http://www.aber.ac.uk/media/index.html , click "textual analysis" and then "content analysis").

1. Summarize the article.

2. Show how the study reflects three points Babbie makes about content analysis.

3. Assess the quality of the study.

EXERCISE 11.5

Name _____

Assess the General Social Survey (see Appendix 1 in this volume) in terms of Babbie's discussion of the issues important to consider when using existing statistics.

ADDITIONAL INTERNET EXERCISES

1. Access
   http://writing.colostate.edu/guides/research/content/pop2a.cfm and
   click "types of content analysis" on the left. Determine whether
   relational and conceptual analysis link with manifest and latent
   coding. Explain your reasoning.

2. Access http://www.temple.edu/sct/mmc/reliability/ for a question
   and answer session on intercoder reliability in content analysis. What
   is intercoder reliability? Why is it important to assess intercoder
   reliability in content analysis?

3. Access the Media and Communications site at
   http://www.aber.ac.uk/media/Sections/http://www.aber.ac.uk/media/i
   ndex.html and read a few of the articles under headings of interest to
   you. Indicate what they add to what is discussed in the Babbie
   chapter.

Chapter 12

Evaluation Research

OBJECTIVES

1.   Identify the purposes of evaluation research.

2.   Identify three factors influencing the growth of evaluation research.

3.   Define and illustrate social intervention.

4.   Describe why it is important to identify the purpose of an intervention.

5.   Define and illustrate the outcome (or response) variable.

6.   Give three examples of experimental contexts that may influence specific evaluation research studies.

7.   Describe two problems with specifying interventions.

8.   Explain why it is important to define the population of possible subjects for whom the program is appropriate.

9.   Compare the two options for measuring variables.

10.   Provide advice for operationalizing success or failure of an intervention.

11.   Apply the classical experimental design to an evaluation research study.

12.   Define quasi-experimental designs.

13.   Define and illustrate time-series designs.

14.   Define and illustrate nonequivalent control groups designs.

15.   Define and illustrate multiple time-series designs.

16.   Define and illustrate cost-benefit analysis.

17.   Show how evaluation research can be less structured and more qualitative.

18.   Discuss why evaluation research is particularly subject to problems in the actual execution of the research.

19.   Summarize three reasons why the implications of evaluation research are not always put into practice.

20.   Define and illustrate social indicators research.

21.   Define and illustrate computer simulation.

22.   Indicate some ethical considerations in evaluation research.

OUTLINE

1. Topics appropriate to evaluation research

2. Formulating the problem: Issues of measurement
    a. Specifying outcomes
    b. Measuring experimental contexts
    c. Specifying interventions
    d. Specifying the population
    e. New versus existing measures
    f. Operationalizing success/failure

3. Types of evaluation research designs
    a. Experimental designs
    b. Quasi-experimental designs
    c. Qualitative evaluations

4. The social context
    a. Logistical problems
    b. Some ethical issues
    c. Use of research results

5. Social indicators research
    a. The death penalty and deterrence
    b. Computer simulation

6. Ethics and evaluation research

SUMMARY

Evaluation research employs the same research methods described in previous chapters. Its uniqueness lies in its purpose: to evaluate the impact of social interventions. It has experienced increased popularity due to the desire of social scientists to influence the quality of life as well as to increased federal requirements for program evaluation and the availability of research funds to meet those requirements.

Many topics are addressed with evaluation research. Evaluation research is a process of determining whether a social intervention has produced the intended result. Needs assessment studies aim to determine the existence and extent of problems. Cost-benefit studies determine whether the results of a program can be justified by relevant expenses. Monitoring studies provide a steady flow of information about something of interest.

Critically important is a clear specification of both the purpose of the intervention to be evaluated as well as the measurement of the key variables. Researchers commonly specify different aspects of desired outcomes and allow for varying levels of success in meeting program objectives. The context within which the study occurs is critical, and hence these variables must be identified and measured as well. Extraneous variables must also be considered. Evaluation researchers must carefully identify the population of possible subjects for the study and must decide whether to create new measures or use existing measures.

Perhaps the most critical aspect of evaluation research is determining whether the program being studied succeeded or failed. Researchers typically allow for gradations of success or failure and often employ cost-benefit analysis in their determination of success. Measurement decisions are complicated by both practical and political considerations because evaluation researchers must work with the people responsible for the program that is being evaluated.

Some evaluation research studies employ the classical experimental design, in which subjects are randomly assigned to experimental and control groups. However, most evaluation research studies cannot meet the demands of random assignment and hence employ quasi-experiments. Time-series designs examine processes occurring over time. Nonequivalent control groups designs are used when an existing "control" group appears similar to the experimental group. Multiple time-series designs are an improved version of the nonequivalent control groups design and employ more than one time-series analysis. Evaluation research need not always be quantitative; less structured and more qualitative evaluation research studies help provide an in-depth understanding of processes producing the observed results.

In comparison to the traditional designs, evaluation research is prone to several types of problems. Logistical problems emerge because evaluation researchers often lack sufficient control over the design in real-life contexts. Researchers must also deal with reluctant and counterproductive administrators who wish to protect favorite programs. Social interventions frequently raise ethical issues, particularly regarding who receives and does not receive what type of stimulus. Political and ideological issues can also intrude.

Evaluation research is designed to be used to make a difference in the execution of some program. Results of evaluation research studies frequently contradict popular beliefs or personal beliefs of administrators involved. Evaluation researchers themselves often fail to present the implications of their research in a way that nonresearchers can understand. Finally, evaluation research results sometimes conflict with the vested interests of individuals in the project being studied.

The use of social indicators has increased in recent years as a nonexperimental form of evaluation research. The use of such indicators makes it possible to monitor many aspects of social life on a large scale. With computer simulation, evaluation researchers can develop mathematical equations describing the relationships that link social variables to one another. By varying one or more of the variables in the equation, social scientists can examine the implications of specific changes.

Ethical issues enter evaluation research. Such researchers are often under pressure to produce particular results, often from sponsors of the study. It is unethical to let such demands influence the design and execution of the study as well as the analysis of results. Furthermore, slanting results may have drastic negative effects on the people served by the program under study because policies may be developed in response to the slanted results.

TERMS

1. applied research
2. computer simulation
3. cost-benefit analysis
4. evaluation research
5. experimental contexts
6. monitoring studies
7. multiple time-series designs
8. needs assessment studies
9. nonequivalent control group design
10. outcome/response variable
11. program evaluation
12. quasi-experiments
13. social intervention
14. social indicators
15. time-series designs
16. vested interests

MATCHING

_____ 1. Experiments that lack random assignment of subjects to experimental and control groups.

_____ 2. The variable that reflects the result of an intervention.

_____ 3. A type of quasi-experimental design that uses a comparison group rather than a control group to which subjects are randomly assigned.

_____ 4. Measurements that reflect the quality or nature of social life.

_____ 5. Ego involvement in a project that often hampers the implementation of evaluation research results.

_____ 6. A form of applied research that assesses the impact of social interventions.

_____ 7. Research that is intended to have some real-world effect.

_____ 8. Variables external to the experiment itself, yet affecting it.

_____ 9. Developing mathematical equations linking social variables so that predictions can be made on future outcomes

_____ 10. Studies that involve measurements taken over time.

TRUE-FALSE QUESTIONS

T   F   1. Applied research is a form of evaluation research.

T   F   2. Los Angeles city officials provided a steady flow of information about how residents responded to an earthquake, a type of evaluation research known as needs assessment.

T   F   3. The city of Palmetto instituted a public relations campaign to increase residents' awareness of the city's recreational facilities and measured the outcome with frequency of park usage. Park usage is known as the response variable.

T   F   4. Specifying the population in evaluation research involves measuring those aspects of the content of an experiment that are thought to affect the experiment.

T   F   5. Perhaps the most difficult task in evaluation research is identifying the relevant population.

T   F   6. Time-series designs are particularly relevant for determining if an intervention has an effect independent of other factors operating over a time period.

T   F   7. Logistical problems include the meddling of program administrators in executing a predetermined design.

T   F   8. Ethical issues are relatively unimportant in evaluation research.

T   F   9. Creating measurements specifically for a study can offer greater relevance and validity than using existing measures.

T   F   10. Experimental contexts refer to factors outside the experiment itself but that may affect it.

REVIEW QUESTIONS

1.   Evaluation research employs which method(s)?
     a.   surveys
     b.   experiments
     c.   observation
     d.   existing statistics
     e.   all of the above

2.   Which one of the following is *correct*?
     a.   Evaluation research is a form of program evaluation.
     b.   Evaluation research is a form of applied research.
     c.   Applied research is a form of evaluation research.
     d.   Applied research is a form of program evaluation.
     e.   Program evaluation is a form of survey research.

3.   The recent growth of evaluation research is due primarily to
     a.   increased methodological sophistication.
     b.   the spurt in the number of social scientists.
     c.   increased federal requirements and financial support.
     d.   disagreement over policy implications.
     e.   the decline in quality of services rendered by social service agencies.

4.   Which one of the following is an example of a program intervention?
     a.   a study of dating behavior among college students
     b.   a study of the effect of gender on participation in voluntary associations
     c.   a study of the effect of religiosity on income
     d.   a study of the effect of team teaching on learning
     e.   a study of the effect of organizational size on productivity

5.   The *most* critical problem in evaluation research is
     a.   dealing with ethical issues.
     b.   measuring the outcome.
     c.   developing an adequate sample.
     d.   dealing with political constraints.
     e.   developing appropriate statistical techniques.

6.   Which of the following strategies is particularly useful for measuring the effects of experimental contexts?
a.   multiple regression
b.   multiple indicators
c.   involving clients
d.   blind designs
e.   control groups

7.   The advantage in using measures that have been used previously is that they have
a.   greater conceptual diversity.
b.   known dimensionality.
c.   known degrees of validity and reliability.
d.   greater statistical rigor.
e.   more policy implications.

8.   Quasi-experimental designs differ from "true" experimental designs
a.   by shorter duration and more political constraints.
b.   by more political constraints and more ethical considerations.
c.   by more ethical considerations and lack of random assignment.
d.   by lack of random assignment and lack of a control group.
e.   by lack of a control group and shorter duration.

9.   Professor Gabino wants to do an evaluation research study on the effect of a new treatment for learning disabilities. She has only one group of clients and wants to assess the effects over the course of a couple of months. Which design would be best?
a.   classical
b.   nonequivalent control group
c.   time-series
d.   posttest-only control group design
e.   factorial

10. Professor Yee wants to do an evaluation study of the effects of a patient education program on patient anxiety. He uses one wing in a hospital for the experiment and compares the results with a similar group of patients in a similar wing in another hospital. Which design would be best?
    a. classical
    b. nonequivalent control group
    c. time-series
    d. posttest-only control group design
    e. factorial

11. Cost-benefit analysis is used to assess
    a. the long-term effect on clients.
    b. the cost of the experiment relative to the social scientific knowledge gained.
    c. how much a program costs in relation to what it returns in benefits.
    d. the costs to society relative to the benefits to the clients.
    e. the costs of the program relative to the number of staff hours invested.

12. Evaluation researchers encounter more logistical problems than other researchers because evaluation research
    a. occurs in the context of real life.
    b. takes longer.
    c. is more costly.
    d. has more measurement problems.
    e. examines more variables.

13. The ethical issues of evaluation research center on
    a. the policy implications.
    b. the administrators of the agency.
    c. the nature of the clients.
    d. the nature of the social intervention.
    e. informed consent.

14. Computer simulation is particularly useful for
   a. measuring social indicators.
   b. analyzing many variables.
   c. outlining policy implications.
   d. improving reliability and validity.
   e. predicting outcomes of specific social changes.

15. The program intervention is equivalent to which one of the following in experimental design?
   a. the contexts
   b. the posttest
   c. the pretest
   d. the stimulus
   e. the external validity

16. Time-series designs fit best into which one of the following?
   a. surveys
   b. true experimental designs
   c. quasi-experimental designs
   d. unobtrusive measures
   e. historical and comparative research

17. The most effective evaluation research is one that
   a. is strictly quantitative.
   b. uses only random samples.
   c. is strictly qualitative.
   d. uses federal support.
   e. is both quantitative and qualitative.

18. Faith did an evaluation research study on the impact of a conflict resolution program on negative social interactions among junior high school students. But teachers and the principal repeatedly interfered both with her placement of students in two groups for comparison and with the length of the program itself. Her study fell victim to
   a. ethical violations.
   b. logistical problems.
   c. time-lagged interventions.
   d. poor measurement.
   e. cost-benefit analysis.

19.   The implications of evaluation research are not always put into practice because
   a.   the implications are not presented in a manner understood by nonresearchers.
   b.   the results contradict deeply held beliefs.
   c.   vested interests come into play.
   d.   of all three factors noted in a, b, and c.
   e.   of just factors noted in a and c.

20.   Bambi wants to do a study on the social conditions in her city by looking at education rates, unemployment, crime rates, and how these intersect to affect the quality of life. Best to use would be
   a.   time-series analysis.
   b.   multiple time-series designs.
   c.   cost-benefit analysis.
   d.   social indicators.
   e.   administrative control.

DISCUSSION QUESTIONS

1. Discuss why measurement is important in evaluation research. Describe the role of measurement in each of the factors Babbie discusses: specifying outcomes, measuring experimental contexts, specifying interventions, specifying the population, new versus existing measures, and operationalizing success/failure.

2. Explain why evaluation researchers find it difficult to employ the classical experimental design, which is best suited to establishing causality. Which designs are more typically used? Why?

3. What are the major reasons why evaluation research results are often ignored? What could be done to remedy this situation?

4. Discuss some of the ethical problems that are particularly troublesome in doing evaluation research. Examine strategies for addressing these problems.

5. Economists have developed a substantial number of indicators to monitor the country's economic health (consumer price index, unemployment rate, and others), many of which are a part of the daily news. How would you explain the scarcity of noneconomic social indicators in the coverage of social issues?

EXERCISE 12.1

Name _____

The purpose of this workbook is to give you some firsthand experience in applying the various aspects of social research, on the assumption that you will learn research methods better that way. The purpose of this workbook, then, is to assist you in understanding the various aspects of social research. In this exercise, you are to design an evaluation study to determine the effectiveness of this workbook. You are planning your project now for implementation next year. So you have some control over the setting in creating your design.

Describe all aspects of your study design, including the likely conclusions.

2. Give specific information regarding the following types of variables: outcomes, experimental contexts, intervention, population, new versus existing measures, and determining success/failure. How will you address and/or measure each?

Outcomes:

Experimental contexts:

Intervention:

Population:

(continued)

New versus existing measures:

Success or failure:

2. Describe which design you selected and why. Outline the study in detail using this design.

3. Describe possible logistical and ethical problems and how you would handle them.

4. Present some fake data and interpret the results.

EXERCISE 12.2

Name _____

You are to design an evaluation research study. Find a newspaper (see www.newspaperlinks.com for links to newspapers online). Find a report of some innovative social program, such as prison reform, a program to provide jobs to young people during school vacations, or a proposal for a new welfare system. Or simply select an innovative social program you are interested in if your instructor so advises. Describe how you would evaluate the success of the program.

1.   Briefly describe the innovative program.

2.   Specify the key dependent variable that is influenced by the program.

3.   Describe as specifically as possible how you would measure that dependent variable.

(continued)

4.   How would you design and conduct your evaluation of the program? Be sure to describe any ways in which the program itself would have to be modified in order to permit your evaluation. Also, specify the data you would need to collect.

5.   Write a hypothetical research report, reporting the "results" of your evaluation of the program. Make up data, and be sure to give your conclusion about the program's success or failure, along with reasons for this conclusion.

EXERCISE 12.3

Name _____

You have been asked to develop a composite measure of the quality of life on your campus.

1. Outline the social indicators you would use to develop such a measure. Be specific. Explain why you selected these indicators.

2. Now suppose data on your social indicators have been collected over a ten-year period. What might you be able to do with these data?

EXERCISE 12.4

Name _____

Visit one of the following Web sites and locate a topic of interest:

> www.lib.umich.edu/govdocs/stsoc.html
> www.whitehouse.gov/fsbr/ssbr.html

1. Describe how the statistics reported on your topic reflect social indicators research.

2. What can you conclude about your topic? If the data span several years, note any changes over time.

ADDITIONAL INTERNET EXERCISES

1. Access the American Evaluation Association's website at http://www.eval.org/ and click "Publications." Then click "Guiding Principles for Evaluators." Describe the major categories used in the guidelines. What obstacles might you encounter in following these guidelines while trying to do a valid and reliable evaluation research study? How do these guidelines link to human subjects issues?

2. Access the Bureau of Justice Assistance's Center for Program Evaluation Web site at http://www.ojp.usdoj.gov/BJA/evaluation/http://www.bja.evaluation website.org/ and click "Program Areas." Select one program area and use specific examples from the description to illustrate points made in the text's chapter on evaluation research.

3. Visit the Harvard Family Research Project site at http://www.hfrp.org and select one of the listed areas under "Research Areas." Describe the selected project and explain why the selected project is an applied research project.

Part 4

Analysis of Data

Chapter 13

Qualitative Data Analysis

OBJECTIVES

1. Define and illustrate qualitative analysis.

2. Compare the connection between data analysis and theory in both qualitative research and quantitative research.

3. Illustrate these ways of looking for patterns in a particular research topic: frequencies, magnitudes, structures, processes, causes, and consequences.

4. Compare the two strategies of cross-case analysis: variable-oriented analysis and case-oriented analysis.

5. Describe the four stages of the constant comparative method used in the grounded theory method.

6. Define and illustrate semiotics.

7. Define and illustrate conversation analysis.

8. Show how coding works in qualitative analysis and compare open coding, axial coding, and selective coding.

9. Explain why standardization is a key principle in quantitative analysis but not so in qualitative analysis.

10. Summarize the role of memoing in qualitative data analysis and compare these types: code notes, theoretical notes, and operational notes.

11. Explain the role of concept mapping.

12. Show how computers can be used in qualitative data analysis.

13. Illustrate how the qualitative analysis of quantitative data is useful.

14. Show how the quality of qualitative research can be evaluated.

15. Indicate relevant ethical issues in qualitative data analysis.

OUTLINE

1.   Linking theory and analysis
     a.   Discovering patterns
     b.   Grounded theory method
     c.   Semiotics
     d.   Conversation analysis

2.   Qualitative data processing
     a.   Coding
     b.   Memoing
     c.   Concept mapping

3.   Computer programs for qualitative data
     a.   QDA programs
     b.   Leviticus as seen through HyperResearch
     c.   Using NVivo to understand women film directors

4.   The qualitative analysis of quantitative data

5.   Evaluating the quality of qualitative research.

6.   Ethics and qualitative data analysis

SUMMARY

Quantitative analysis of social research data is the most dominant approach in the social sciences. Its emphasis on statistical analysis, however, sometimes conceals another very useful approach to making sense of social observations: qualitative analysis. Qualitative data analysis include methods for examining and making sense of social research data without converting them to a numerical format. It is easy to lose sight of theory in quantitative data analysis. In qualitative data analysis, however, data collection, analysis, and theory are more closely connected.

The approach to theory used in qualitative data analysis can best be described as a process of looking for plausible relationships proposed among concepts and sets of concepts. Discovering patterns lies at the root of this explanatory focus, which can be done by looking at different analytical topics. Frequencies simply measure how often something occurs. Magnitudes reflect the levels of particular variables. Structures reflect different types of a variable. Processes reflect an order among the elements of structure. Causes are the factors influencing something of interest. Consequences are the outcomes of a particular variable.

Qualitative data analysts look for patterns appearing across several observations that usually represent different cases under study, known as cross-case analysis. Two types of cross-case analysis can be used. Variable-oriented analysis examines the impact of several independent variables on a particular dependent variable across many cases. Case-oriented analysis, on the other hand, involves looking deeply at one or just a few cases. The researcher may see the critical elements of one subject's experiences as instances of more general social concepts or variables.

The Grounded Theory Method stresses the importance of establishing theories on an inductive basis. This approach begins with observations instead of hypotheses and discovers patterns and develops theories from the ground up, with no preconceptions. The Grounded Theory Method uses the constant comparative method, which includes four stages. The first compares incidents applicable to each category. The second integrates categories and their properties. The third delimits the theory, and the fourth writes the theory.

Semiotics is the science of signs and addresses symbols and meanings. Those following this approach stress that meanings attached to signs are socially constructed. Signs are any things that are assigned special meanings. Conversation analysis examines closely the ways people converse with each other in order to uncover the implicit assumptions and structures in social life. It is based on three assumptions. First, conversation is a socially structured activity. Second, conversations must be understood contextually. Third, detailed transcriptions of conversations are required to understand the meaning and structure of conversation.

Qualitative data processing brings order to qualitative data. Coding is central and involves classifying or categorizing pieces of data. Coding in qualitative data processing centers on the concept as the organizing principle. This differs from the emphasis on a standardized unit of analysis in more statistical analyses. Qualitative data analysis involves open coding, the naming and categorizing of phenomena through close examination of data. This important first step yields clues for identifying similarities and differences and helps uncover people's assumptions. Axial coding is a second type of coding which reanalyzes the results of open coding to identify the core concepts. Finally, selective coding seeks to identify the central code in the study, the one that the other codes all relate to.

As qualitative data analysts code data, they often use the technique of memoing– writing notes to yourself and others involved in the project. Three types of memos are used in the Grounded Theory Method. Code notes identify the code labels and their meanings. Theoretical notes address the meanings of concepts, relationships among concepts, and theoretical propositions. Operational notes address methodological issues, such as data collection circumstances. Some-times qualitative data analysts portray their concepts in a graphical format, known as concept mapping, which helps identify relationships among concepts more clearly. Various computer programs have greatly simplified qualitative data analysis.

Qualitative data analysis of quantitative data often yields new insights and shows how the two approaches are not incompatible. Judging the quality of qualitative research is not as easy as judging the quality of quantitative research.

Yet similar standards exist. It is important to assess the validity and reliability of qualitative research, using similar principles as used in quantitative research. Additional issues should also be addressed, such as credibility of results, utility, inferences drawn, contexts of data sources, diversity of perspectives, and theoretical assumptions. Of course, other issues apply regardless of the type of data analysis, such as quality of research design and execution, consistency between results and conclusions, and quality of samples.

Qualitative data analysis also carries ethical considerations. Reliance on researchers' subjective judgments heightens concerns about researcher bias, although strategies exist to reduce this source of error. In addition, protecting subjects' privacy is paramount because qualitative researchers usually know their subjects.

TERMS

1. axial coding
2. case-oriented analysis
3. code notes
4. coding
5. concept mapping
6. constant comparative method
7. conversation analysis
8. cross-case analysis
9. Grounded Theory Method
10. memoing
11. open coding
12. operational notes
13. qualitative analysis
14. selective coding
15. semiotics
16. theoretical notes
17. variable-oriented analysis

MATCHING

1. The type of cross-case analysis that focuses on a particular variable.

_____

2. A method for building theories on an inductive basis that begins
with observations rather than hypotheses and seeks to discover patterns.

_____

3. The process of classifying or categorizing individual pieces of data.

_____

4. A type of memoing in which the researcher reflects on the meanings of concepts, relationships among concepts, and theoretical propositions.

_____

5. A form of coding that aims to identify the core concepts in a study.

_____

6. Putting concepts in a graphical format to enable the researcher to think out the relation-ships among concepts more clearly.

_____

7. Methods for examining social research data without converting them to a numerical format.

_____

8. The type of cross-case analysis that involves looking more closely into a few particular cases.

_____

9. A form of coding that seeks to identify the central code in a study, the one that the other codes all relate to.

_____

10. A form of coding that provides the initial classification and labeling of concepts in qualitative data analysis.

TRUE-FALSE QUESTIONS

T   F   1. Case-oriented analysis is similar to idiographic explanation.

T   F   2. The constant comparative method is employed by the Grounded Theory Method.

T   F   3. Samantha wishes to study the sociological significance of jacket insignia. Best to use would be conversation analysis.

T   F   4. Conversation analysis lies at the heart of ethnomethodology.

T   F   5. The part of qualitative analysis that pertains to the naming and categorizing of phenomena through close examination of data is known as closed coding.

T   · F   6. Memoing in data coding is the writing of notes to yourself and others involved in a project.

T   F   7. Concept mapping includes locating our concepts within the larger literature on the topic.

T   F   8. Qualitative analysis cannot be applied to quantitative data.

T   F   9. Axial coding aims to identify the core concepts in a study whereas selective coding seeks to identify the central code to which the other codes are all related to.

T   F   10. The number of computer programs available for qualitative data analysis has increased considerably in recent years.

REVIEW QUESTIONS

1. Professor Martini has done a qualitative analysis of fraternity parties. He selected one party for intensive analysis of the social interactions that occur between women and men. He is using which method?
   a.   variable-oriented analysis
   b.   memoing
   c.   open coding
   d.   case-oriented analysis
   e.   operational notes

2. Martini then compared incidents at various parties applicable to his categories of analysis, integrated the categories and their properties, delimited the theory, and wrote the theory. Which method is he using?
   a.   constant comparative method
   b.   semiotics
   c.   coding
   d.   memoing
   e.   sorting memo

3. April spent some time in the student union listening to the discussions students had while eating lunch. She wanted to see how such discussions are socially constructed and how they vary by the social context, such as the size of the group. She was using
   a. semiotics.
   b. the constant comparative method.
   c. the Grounded Theory method.
   d. variable-oriented analysis.
   e. conversation analysis.

4. June did a study of the meanings attached to the logos on shirts and jackets. She was using
   a. elemental memos.
   b. code notes.
   c. conversation analysis.
   d. case-oriented analysis.
   e. semiotics.

5. Instead of the standardized unit of analysis in quantitative analysis, qualitative analysis uses which one of the following as the organizing principle for coding?
   a. the concept
   b. the variable
   c. the sorting memo
   d. concept mapping
   e. semiotics

6. Before doing qualitative data analysis, an important first step is to name and categorize phenomena through close examination of the data. This process is known as
   a. theoretical notes.
   b. open coding
   c. concept mapping.
   d. qualitative analysis
   e. memoing

7. Which type of notes in the memoing process deal with methodological issues?
   a.   code notes
   b.   theoretical notes
   c.   operational notes
   d.   elemental notes
   e.   integrating notes

8. Professor Idiazinni has completed her qualitative study of Girls Scouts campouts and is now engaged in coding to identify the core concepts in her study. She is employing
   a.   open coding
   b.   axial coding
   c.   selective coding
   d.   ethnographic coding
   e.   semiotic coding

9. Penelope was having difficulty seeing the relationships among the concepts in her qualitative analysis of teacher-student interactions in the classroom. She decided to put her concepts in a graphical format on a large piece of paper. She is using
   a.   coding.
   b.   code notes.
   c.   sorting memos.
   d.   the constant comparative method.
   e.   concept mapping.

10.   Which one of the following statements is correct?
   a.   Theory is closely connected to data analysis in both quantitative and qualitative analysis.
   b.   Theory is not closely connected to data analysis in either quantitative or qualitative analysis.
   c.   Data collection, analysis, and theory are more intimately intertwined in quantitative analysis than in qualitative analysis.
   d.   Data collection, analysis, and theory are more intimately intertwined in qualitative analysis than in quantitative analysis.
   e.   None of these statements is true.

11. Professor Cessna identified how often students cheat and the severity or level of the cheating. She was looking at
    a.   frequencies and structures.
    b.   structures and causes.
    c.   consequences and causes.
    d.   processes and magnitudes.
    e.   frequencies and magnitudes.

12. Cessna then looked at the different types of cheating as well as how cheating affected the student and the professor. She now was looking at
    a.   frequencies and magnitudes.
    b.   structures and consequences.
    c.   causes and consequences.
    d.   processes and causes.
    e.   processes and consequences.

13. Professor Idiazinni has completed her qualitative study of Girls Scouts campouts and now wishes to move beyond identifying the core concepts in her study to identifying the central code to which all the other codes are related. She should use
    a.   open coding.
    b.   axial coding.
    c.   selective coding.
    d.   ethnographic coding.
    e.   semiotic coding.

14. Professor Nguyen studied which factors helped explain why gangs emerged more frequently in some neighborhoods than others. Which way of looking for patterns did he employ?
    a.   structures
    b.   processes
    c.   causes
    d.   consequences
    e.   frequencies

15. Shanique sought to uncover the different types of playground interactions among children in her qualitative study. Which way of looking for patterns did she employ?
    a.   structures
    b.   processes
    c.   causes
    d.   consequences
    e.   frequencies

16. After uncovering the different types of playground interactions among children in her qualitative study, Shanique sought to uncover any time order among the various types of social interactions that occurred. Which way of looking for patterns did she employ?
    a.   structures
    b.   processes
    c.   causes
    d.   consequences
    e.   frequencies

17. The Grounded Theory Method employs
    a.   semiotics
    b.   conversation analysis
    c.   case-oriented analysis
    d.   variable-oriented analysis
    e.   constant comparative method

18. The key process in the analysis of qualitative data is
    a.   surveys.
    b.   deduction.
    c.   induction.
    d.   coding.
    e.   grounded theory method.

19. The approach that says that meanings reside in minds is
    a.   semiotics.
    b.   conversation analysis.
    c.   constant comparative method.
    d.   variable-oriented analysis.
    e.   case-oriented analysis.

20. Ethics in qualitative data analysis particularly involve
    a.   potential researcher bias and harm to subjects.
    b.   harm to subjects and researcher integrity.
    c.   researcher integrity and meeting Institutional Research Boards' demands.
    d.   meeting Institutional Research Boards' demands and protecting subjects' privacy.
    e.   protecting subjects' privacy and potential researcher bias.

DISCUSSION QUESTIONS

1. Give examples of the different ways of looking for patterns in a particular research topic.

2. Show how a given topic of research could be analyzed differently with the two methods of cross-case analysis: variable-oriented analysis and case-oriented analysis.

3. Rewrite the stages in the constant comparative method as used in the Grounded Theory Method.

4. Provide advice for performing qualitative data processing.

EXERCISE 13.1

Name _____

Exercise 10.1 involved data gathering for a study of the social dynamics of jaywalking. Review your field notes if you have completed that exercise. Complete the exercise now if you have not done so before.

1. Use the strategies outlined in this chapter for processing the data. Indicate which strategies you have used and summarize the results of the data processing. Hold off with analyzing the results until the next question.

2. Use the strategies outlined in this chapter for analyzing the data. Indicate which strategies you used and summarize the results of the data analysis.

EXERCISE 13.2   Name _____

Exercise 10.2 involved data gathering for a study of the social interactions among strangers. Review your field notes if you have completed that exercise. Complete the exercise now if you have not done so before.

1. Use the strategies outlined in this chapter for processing the data. Indicate which strategies you have used and summarize the results of the data processing. Hold off with analyzing the results until the next question.

2. Use the strategies outlined in this chapter for analyzing the data. Indicate which strategies you used and summarize the results of the data analysis.

EXERCISE 13.3

Name _____

Use semiotics to analyze signs or symbols that strike you as having social scientific interest. Examples might include the messages and insignias on shirts and jackets, billboards, popular music, names given to houses where students live, and body piercing and tattoos. You are not limited to these examples.

1. Describe specifically which symbols you studied, how many, and why you selected these symbols.

2. Use the strategies outlined in this chapter for processing the data. Indicate which strategies you have used and summarize the results of the data processing. Hold off with analyzing the results until the next question.

3. Use the strategies outlined in this chapter for analyzing the data. Indicate which strategies you used and summarize the results of the data analysis.

ADDITIONAL INTERNET EXERCISES

1.   Visit http://www.udel.edu/chem/white/teaching/ConceptMap.html and describe the steps in concept mapping. Apply the steps to an area of interest.

3.   Visit http://ebn.bmj.com/cgi/content/full/3/3/68 for a review of qualitative data analysis in nursing research. Apply the principles discussed to a different area of qualitative research.

4.   Visit http://carbon.cudenver.edu/~mryder/itc_data/semiotics.html and select one source about semiotics to review and summarize.

Chapter 14

Quantitative Data Analysis

OBJECTIVES

1. Define quantitative analysis.

2. Describe the coding process and compare two approaches to developing code categories.

3. Offer some advice for coding a set of data.

4. Identify two functions of codebooks.

5. Present some of the common elements in codebook formats.

6. Describe several ways for entering data.

7. Define and give examples of univariate analysis.

8. Define and explain the utility of frequency distributions and marginals.

9. Explain how to calculate percentages.

10. Define central tendency.

11. Compare the mode, mean, and median in terms of calculation and interpretation.

12. Describe the information provided by measures of dispersion.

13. Compare the range, standard deviation, and interquartile range in terms of calculation and interpretation.

14. Distinguish continuous variables from discrete variables by definition and example.

15. Provide guidelines for balancing the demands of detail versus manageability of data presentation.

16. Differentiate the goals of univariate, bivariate, and multivariate analyses.

17. Identify the goal of subgroup comparisons.

18. Explain two techniques for handling the "don't know" response.

19. Differentiate dependent variable from independent variable by definition and example.

20. Describe how bivariate tables are presented and analyzed.

21. Explain how multivariate tables are presented and analyzed.

22. Indicate relevant ethical issues in quantitative analysis.

OUTLINE

1. Quantification of data
    a, Developing code categories
    b. Codebook construction
    c. Data entry

2. Univariate analysis
    a. Distributions
    b. Central tendency

      c.   Dispersion
      d.   Continuous and discrete variables
      e.   Detail versus accountability

3.   Subgroup comparison
      a.   "Collapsing" response categories
      b.   Handling "don't knows"
      c.   Numerical descriptions in qualitative research

4.   Bivariate analysis
      a.   Percentaging a table
      b.   Constructing and reading bivariate tables

5.   Introduction to multivariate analysis

6.   Sociological diagnostics

7.   Ethics and quantitative data analysis

SUMMARY

Quantitative analysis employs techniques whereby researchers convert data to a numerical form and subject it to statistical analysis. Quantification is the process of converting data to a numerical format, which must be in a machine-readable form so that the data can be read and manipulated by computers. The process for coding closed-ended questions is quite straightforward. Responses to open-ended questions are nonnumerical and must be coded before analysis. In the process, we reduce a wide variety of idiosyncratic items of information to a more limited set of attributes composing a variable. Remember to code as much detail as possible because code categories can later be combined if the analysis desired does not require the originally coded level of detail.

The coding scheme may already be established before the data are examined, or it may be developed after examining a number of cases to determine what set of code categories would be most appropriate. Coding choices should reflect research purposes and should reflect the logic that emerges from the data. Code categories should be both exhaustive (covering all possibilities) and mutually exclusive (no overlapping categories). It is important to verify reliability of coding procedures.

A codebook is a document that describes the locations of variables and lists the codes assigned to the attributes of the variables. It is the primary guide used in the coding process and serves as a guide for locating variables and interpreting the attributes of variables. Each variable is identified with an abbreviated name. The full definition of the variable is also reported. The codebook will also show the attributes for each variable. The codebook helps the researcher to decide which variables should be examined in which fashion.

Various options exist for converting data into a machine-readable format. Coding could be done directly on the questionnaire itself. Optical scan sheets could be used. Data entry sometimes occurs during the process of data collection, such as with computer assisted interviewing strategies. Online surveys enables respondents to enter their own data.

Social scientists perform a variety of analysis strategies on their data. There are three basic strategies: univariate analysis, bivariate analysis, and multivariate analysis. Univariate analysis involves the examination of only one variable at a time. Data are organized into frequency distributions, which are listings of the number of cases in each attribute of a variable. Each case may be listed separately, or they may be grouped in the form of marginals. Marginals sacrifice some detail in the data but provide a more manageable format for describing a distribution. In addition to raw numbers, percentages may be presented by dividing the number of cases per attribute by the base. Careful attention must be paid to determining the base, particularly in the case of missing data.

The detailed information in raw data may be simplified by calculating a measure of central tendency, a single value that summarizes a distribution. The mode reflects the attribute with the greatest frequency; the mean is calculated by summing the values and dividing by the number of cases; and the median reflects the value that cuts the distribution in half.

Measures of dispersion are calculated to reflect the spread of scores. The range is the distance between the highest and lowest values; the standard deviation reflects the difference between each score and the mean; and the interquartile range reflects the range after removing the top fourth and bottom fourth of the scores.

Which measure of dispersion or central tendency to use is determined by the nature of the data and the purpose of the analysis. For example, continuous variables and discrete variables are treated differently. Researchers sometimes violate some of the statistical assumptions, and this is legitimate to the extent that it helps interpret the data. But be aware that such procedures may lead you to think that the results represent something truly precise. Researchers are also constrained by the conflicting goals of providing complete detail versus presenting data in a manageable format.

Subgroup comparisons yield a descriptive analysis to two variables by presenting descriptive univariate data for each of several subgroups. Occasionally categories must be collapsed to enhance the analysis. "Don't know" responses can be included or excluded, depending on the purpose of the study. Subgroup comparisons focus on describing the people (or other units of analysis), and bivariate analysis focuses on the relationship between the variables themselves. Hence bivariate analysis assumes an explanatory goal.

The explanatory goal of bivariate analysis involves making causal assertions regarding a dependent variable that is at least partially determined by an independent variable. Hence bivariate analysis tables should be consistent with the logic of independent and dependent variables. Typically the attributes of the independent variable are used as column headings, and the attributes of the dependent variables are used as row headings. Percentages are then computed down (adding up to 100 percent) and compared across. But these conventions are sometimes reversed. Regardless of how a bivariate analysis is presented, it is critical to interpret such a table by comparing the independent variable subgroups with one another in terms of the attributes of the dependent variable.

Several guidelines should be followed in presenting and interpreting bivariate tables (also known as contingency tables): 1) the contents of the table should be described in a title; 2) the original content of the variables should be clearly presented in either the table itself or the text; 3) the attributes of each variable should be clearly described; 4) the base on which percentages are computed should be shown; and 5) the table should indicate the number of cases omitted from the analysis.

The logic and procedures in bivariate tabular analysis also apply to multivariate tables. But instead of explaining a dependent variable on the basis of one independent variable, such analysis explains a dependent variable on the basis of multiple independent variables. Both the separate and the combined effects of the independent variables on the dependent variable can then be examined. When the dependent variable is dichotomous, the multivariate table can be simplified by omitting the data for one attribute of the dependent variable. The two independent variables are then typically placed on the top and the side of a table, and the percentages recorded reflect one attribute of the dependent variable for each combination of attributes of the independent variables. The base for each cell should be reported.

Ethics enters into quantitative analysis as well. Quantitative analysis must guard against defining and measuring variables in ways that encourage one finding over another. Analysts should report formal hypotheses as well as more informal expectations that did not pan out. Subjects' privacy rights need to be respected.

TERMS

1. base
2. bivariate analysis
3. central tendency
4. codebook
5. coding
6. contingency table
7. continuous variable
8. dependent variable
9. discrete variable
10. dispersion
11. frequency distribution
12. independent variable
13. interquartile range
14. marginals
15. mean
16. median
17. mode
18. multivariate analysis
19. percentages
20. quantitative analysis
21. range
22. standard deviation
23. subgroup comparisons
24. univariate analysis

MATCHING

_____ 1. The examination of the effects of two or more independent variables on a dependent variable.

_____ 2. A measure of central tendency that measures the most frequent attribute.

_____ 3. The measure of central tendency that is the arithmetic average.

_____ 4. The measure of central tendency that divides the distribution in half.

_____ 5. The measure of dispersion that resembles the standard error of a sampling distribution and that is interpreted in terms of the normal curve.

_____ 6. A method of univariate analysis that simply reports the number of cases in each attribute category.

_____ 7. A variable that influences another variable.

_____ 8. A variable that is influenced by another variable.

_____ 9. A document that describes the locations of variables and lists the assignments of codes to the attributes composing those variables.

_____ 10. Refers to the way values are distributed around some central value, such as an average.

TRUE-FALSE QUESTIONS

T   F   1. There is only one way for developing code categories: begin with a relatively well-developed coding scheme.

T   F   2. A research manual is a document that describes the locations of variables and lists the assignment of codes to the attributes composing those variables.

T   F   3. When the total number of subjects is an even number, you cannot calculate a median.

T   F   4. Which measure of central tendency to use depends on the nature of your data and the purpose of your analysis.

T   F   5. The standard deviation represents the distance between the highest and lowest values.

T   F   6. Gender is an example of a discrete variable.

T   F   7. Tables should be percentaged within the categories of the independent in terms of the dependent variable.

T   F   8. Multivariate analysis is particularly useful for sociological diagnostics.

T   F   9. A continuous variable increases steadily in tiny fractions.

T   F   10. The logic of multivariate analysis is simply an extension of bivariate analysis.

REVIEW QUESTIONS

1. The end product of quantifying data is to convert measurements into
   a. scales.
   b. typologies.
   c. machine-readable forms.
   d. sophisticated statistics.
   e. bytes.

2. Which one of the following is an example of a frequency distribution of grouped data for age?
   a. 18-25, 40 people; 26-50, 60 people; over 50, 20 people
   b. 18, 10 people; 19, 11 people; 20, 26 people; 21, 15 people
   c. 26, 28, 28, 30, 31, 33, 35, 41, 41, 50
   d. the mean age is 45.1
   e. the median age is 38.6

3. Frequency distributions of grouped data are often referred to as
   a. indexicals.
   b. dispersion.
   c. epsilon.
   d. monotonicity.
   e. marginals.

4. The process of assigning numbers to questionnaire responses is known as
    a. memoing.
    b. substitution.
    c. indexing.
    d. percentaging.
    e. coding.

5. Which measure of central tendency divides the distribution into two halves?
    a. dispersion
    b. mode
    c. marginal
    d. median
    e. mean

6. With several extreme scores, which measure of dispersion would be most appropriate?
    a. range
    b. standard deviation
    c. standard error
    d. confidence level
    e. interquartile range

7. A fundamental dilemma in presenting data is
    a. scope versus adequacy.
    b. nominal versus ordinal values.
    c. univariate versus bivariate.
    d. detail versus manageability.
    e. variation versus central tendency.

8. Descriptive analyses are most often accomplished with which type of statistical analysis?
    a. bivariate
    b. univariate
    c. dispersion
    d. multivariate
    e. interpretation

9.   Comparing the church attendance of men and women for descriptive purposes is an example of
    a.   univariate analysis.
    b.   multivariate analysis.
    c.   subgroup comparisons.
    d.   explanatory analysis.
    e.   bivariate analysis.

10.   Professor Dornby analyzed the relationship between age and voting behavior. This is an example of which type of analysis?
    a.   univariate.
    b.   bivariate.
    c.   multivariate.
    d.   descriptive.
    e.   ex post facto.

11.   The fundamental rule in reading tables is to
    a.   percentage down.
    b.   percentage across.
    c.   read across the categories of the dependent variable in analyzing the independent variable.
    d.   read across along the categories of the independent variable in analyzing the dependent variable.
    e.   percentage and read in terms of the total number of cases.

12.   Another term for bivariate tables is
    a.   dependency tables.
    b.   independency tables.
    c.   marginals tables.
    d.   cross-sectional tables.
    e.   contingency tables.

13.   Professor Goldsmid studied the relationship between gender and religiosity while controlling for social class. This is an example of which type of analysis?
    a.   univariate
    b.   bivariate
    c.   multivariate analysis
    d.   descriptive
    e.   ex post facto

14. Combining two independent variables and one dependent variable into one table is possible *only* when
    a. the dependent variable is dichotomous.
    b. the independent variables are dichotomous.
    c. both the independent and dependent variables are dichotomous.
    d. a test of statistical significance is applied.
    e. a bivariate analysis is done.

15. Sociological diagnostics are used to
    a. diagnose what is wrong with a contingency table.
    b. diagnose the weaknesses in a measure.
    c. spiff up the statistics in research reports.
    d. help determine the current state of affairs and point the way to where we want to go.
    e. tease out the ethical implications in quantitative data analysis.

16. In reading a table that someone else has constructed, the first thing you need to determine in order to interpret it correctly is
    a. find out in which direction it has been percentaged.
    b. whether multivariate analysis was employed.
    c. how big the sample is.
    d. how the variables were measured.
    e. what the standard deviation is.

17. A good suggestion for handling "don't know" responses in your tables is to
    a. ignore them.
    b. always delete them.
    c. always include them.
    d. report your data with and without the "don't knows."
    e. interview some of them to find out why they said "don't know."

18.   An effective strategy for handling response categories with few cases is to
        a.   expand at least some of the response categories.
        b.   collapse at least some of the response categories.
        c.   calculate the mean for the largest categories.
        d.   survey additional respondents.
        e.   sit down and first discuss the ethical considerations.

19.   Which measure of central tendency would a home builder be most interested in?
        a.   median
        b.   standard deviation
        c.   mode
        d.   interquartile range
        e.   mean

20.   The two basic strategies to choose from in coding are
        a.   relax–the codes will appear or else use qualitative reasoning.
        b.   use qualitative reasoning or uncover latent content.
        c.   uncover latent content or uncover manifest content.
        d.   uncover manifest content or begin with a well developed coding scheme.
        e.   begin with a well developed coding scheme or generate codes from the data.

DISCUSSION QUESTIONS

1. Compare univariate and bivariate analyses in terms of definition, purpose, advantages, and limitations.

2. Compare the three measures of central tendency (mode, median, mean) in terms of definition, calculation, interpretation, and limitations.

3. Subgroup comparison is considered to be somewhat intermediate between univariate and bivariate analyses. In what sense does subgroup analysis serve a descriptive purpose, and in what sense does it serve an explanatory purpose?

4. Explain the logic in constructing and interpreting bivariate tables. Why should you percentage within categories of the independent variable and then compare across the categories of the independent variable in terms of the dependent variable?

EXERCISE 14.1

Name _____

You are to code the responses you got in answer to Question 7 of the questionnaire used in Exercise 9.3.

1.   Code each of the responses as:
        (1)   favorable to capital punishment,
        (2)   unfavorable to capital punishment, or
        (3)   neutral to capital punishment

RESPONSE   CODE (1,2,3)

1.   _____

2.   _____

3.   _____

4.   _____

5.   _____

6.   _____

7.   _____

(continued)

8.   _____

9.   _____

10.   _____

2.   Describe any difficulties you experienced in coding and note how you resolved them. Explain how you decided to apply codes of 1, 2, or 3.

EXERCISE 14.2

Name _____

The number of movies attended in the preceding three months are given below for 30 people. Calculate the mean, the median, the mode, and the range. Show your work. Interpret each statistic.

0	15	26	8	2
3	11	13	0	3
0	1	17	1	9
7	4	18	12	10
10	4	2	7	2
2	5	9	6	6

1. Calculate the mean and interpret the results.

2. Calculate the median and interpret the results.

3. Calculate the mode and interpret the results.

4. Compare the three measures of central tendency you have calculated in terms of the information provided.

5. Calculate the range and interpret the results.

EXERCISE 14.3

Name _____

Presented below are some hypothetical data representing 50 people: (1) gender (M=male, F=female) and (2) whether they attended church last week (Y=yes, N=no).

In the steps below you will be asked to construct the appropriate bivariate percentage table showing the relationship between gender and church attendance and to give your interpretation of that table.

M-N F-N   M-N M-Y F-N   F-Y   F-Y   M-N M-Y
M-Y F-Y   M-N M-Y F-Y   F-N   M-N M-N F-N   F-Y
M-N F-Y   M-N M-Y F-Y   F-Y   M-N M-N F-Y   M-N
F-Y   M-N M-Y F-Y   F-N   F-Y   M-Y F-Y   M-N M-N
F-Y   F-Y   M-N F-N   M-N F-Y   M-N F-Y   F-Y   M-N

1.   Construct the bivariate percentage table appropriate for examining the relationship between sex and church attendance. Be sure to follow Babbie's suggestions for setting up a table correctly.

2.   Interpret the table. That is, compare the percentages on the independent variable in terms of the categories of the independent variable.

EXERCISE 14.4

Name _____

Presented below is a hypothetical trivariate percentage table.

Percentage in Favor of
Gun Control

	Men	Women
Young	50%	70%
Old	70%	90%

1.   Explain the logical structure of the table.

2.   Give your interpretation of the table.

EXERCISE 14.5

Name _____

Babbie discusses the usefulness of collapsing response categories to facilitate analyses. Use SPSS or another data analysis program as indicated by your instructor to apply this principle to the variable REGION in the General Social Survey (see Appendix 1 in this volume). Run a frequency distribution on REGION and interpret the results. Be sure to use the "valid percent" category, which omits missing values. Then collapse REGION as follows: "New England" and "Middle Atlantic" into "East;" "East North Central" and "West North Central" into "North Central;" "South Atlantic," "East South Central," and "West South Central" into "South;" and "Mountain" and "Pacific" into "West." Recalculate the percentages.

1. Report the frequency distribution *before* collapsing and summarize the results. Be sure to use the "valid percent" category.

2. Report the frequency distribution *after* collapsing and summarize the results. Be sure to use the "valid percent" category.

3. Compare the results for the two tables and note why the collapsed table may be more useful for analysis purposes.

EXERCISE 14.6

Name _____

Select eight variables that interest you from the General Social Survey codebook in Appendix 1 of this volume. Select several items that could be used as independent variables and several that could be used as dependent variables. Use SPSS or another data analysis program as indicated by your instructor to produce frequency distributions of the variables. Request measures of central tendency.

Interpret the frequency distributions; use the "valid percent" column because it omits missing data. Interpret the appropriate statistics for five of the variables. Your interpretations should mention the level of measurement of each variable and refer only to those statistics appropriate for that level. (Note that the computer will calculate any statistics on your data whether appropriate or not.) Write your interpretations directly on the output (unless your instructor advises otherwise). Finally, write your name on the top page and list the page numbers of the printout containing your analyses.

EXERCISE 14.7

Name _____

Use SPSS or another data analysis program as indicated by your instructor to construct six tables from the variables you analyzed in Exercise 14.6 involving the General Social Survey codebook contained in Appendix 1 of this volume. Select sufficient independent and dependent variables now if you have not done Exercise 14.6.

Select two tables to interpret. Be sure to 1) compare percentages on the dependent variable across categories of the independent variable, 2) summarize the relationship, and 3) briefly explain the relationship by providing possible reasons for why the relationship exists or doesn't exist. (Your task of learning how to read the tables might be more meaningful at this point if you choose variables with relatively few categories. This simplifies the structure of the tables.) Do this directly on the output (unless your instructor advises otherwise). Write your name on the top page and list the page numbers of the printout containing your analyses.

ADDITIONAL INTERNET EXERCISES

1. Visit the Census Bureau site at http://census.gov and click "American Fact Finder" and then "Fact Sheet." Examine the sex, age, and race data for the entire United States and also for two states. Describe the univariate distributions on each variable for each of the three locations. Construct a bivariate table of location by sex.

2. Visit the Census Bureau site at http://census.gov and click "American Fact Finder" and then "People" on the left. Then click the "age and sex" statement. Examine the data for the United States and construct a bivariate table of that data using age categories of under 18, 18 to 64, and 65 and older. What does the table indicate?

3. Access the FedStats homepage at http://www.fedstats.gov/, which is a "gateway to statistics from over 100 federal agencies." This is a fun site to browse. Select a topic of interest and select and report the statistics of interest.

Chapter 15

The Elaboration Model

OBJECTIVES

1. Describe the goal of the elaboration model.

2. Summarize the historical origins of the elaboration model.

3. Define and illustrate control variables.

4. Define partial tables and partial relationships.

5. Differentiate antecedent from intervening variables by definition and example.

6. Differentiate the following outcomes of the elaboration model by definition and example: replication, explanation, interpretation, and specification.

7. Differentiate a suppressor variable from a distorter variable by definition and ex-ample.

8. Discuss the positive and negative features of ex post facto hypothesizing.

OUTLINE

1.   The origins of the elaboration model

2.   The elaboration paradigm
   a. Replication
   b.   Explanation
   c.   Interpretation
   d.   Specification
   e.   Refinements to the paradigm

3.   Subgroup comparison
   a.   "Collapsing" response categories
   b.   Handling "don't knows"
   c.   Numerical descriptions in qualitative research

4.   Elaboration and ex post facto hypothesizing

SUMMARY

The elaboration model involves multivariate analysis to portray more clearly the relationship between two variables by introducing additional control variables. The elaboration model derives historically from Samuel Stouffer's research during World War II. Faced with unexpected findings from studies of soldier morale, Stouffer suggested logical explanations for the findings based on the theoretical concepts of reference group and deprivation. Patricia Kendall and Paul Lazarsfeld later formalized the elaboration model suggested in Stouffer's work by constructing hypothetical tables consistent with Stouffer's explanations.

The elaboration model extends a bivariate relationship by introducing a control variable. The entire sample is divided into the categories of the control variable, and then the original bivariate relationship for each subgroup is recomputed separately. The resulting partial relationships, one for each attribute of the control variable, are then compared with the original bivariate relationship to determine the effect of the control variable on the original relationship.

There are four major outcomes possible in elaboration analysis. Replication occurs whenever the partial relationships are essentially the same as the original bivariate relationship. Hence the original relationship has been replicated under test conditions. Specification occurs when the partial relationships are significantly different from each other. For example, the original bivariate relationship may be found in one subgroup but not in the other. This outcome specifies the conditions under which the relationship occurs or fails to occur.

Before the next two outcomes can be identified, i.e., explanation and interpretation, the control variable must be identified as antecedent to the other two variables or as intervening between them. Explanation occurs when an original bivariate relationship is explained away through the introduction of a control variable, and the control variable is antecedent to both the independent and dependent variables. The partial relationships must be zero or significantly less than those found in the original relationship. Such an explained relationship is often called a spurious relationship. On the other hand, similar findings can be the result of interpretation when the control variable occurs between the independent and dependent variables rather than prior to both variables. The partial relationships must again be zero or significantly less than those found in the original relationship.

Methodologists have extended the elaboration model beyond that formulated by Stouffer and by Kendall and Lazarsfeld. For example, Rosenberg suggested using the elaboration model even when the original relationship is zero. Doing so may uncover a suppressor variable, which conceals the relationship in the original table. The basic model also fails to provide guidelines for specifying what constitutes a significant difference between the original and the partials and fails to consider the possibility that partial relationships might be stronger or the reverse of the original. The latter case is caused by a distorter variable. Finally, the basic approach employs only dichotomous control variables, but nondichotomous control variables can also be used.

Ex post facto hypothesizing occurs when a social scientist observes an empirical relationship and then suggests one or more hypotheses to explain that relationship. On the surface, this procedure violates the principle that hypotheses must be disconfirmable and is, therefore, inappropriate. But the elaboration model often provides the opportunity to suggest and test additional hypotheses regarding a bivariate relationship once the data are already at hand. The acceptance of hypotheses is primarily a function of the extent to which they have been tested by elaboration and not disconfirmed; ex post facto hypothesizing may be a central part of this testing process if it is appropriately applied.

TERMS

1. antecedent variable
2. distorter variable
3. elaboration model
4. explanation
5. ex post facto hypothesizing
6. interpretation
7. intervening variable
8. partial relationships

9. partial relationship
10. replication
11. specification
12. spurious relationship
13. suppressor variable
14. test variable
15. zero-order relationship

MATCHING

_____ 1. The technique used to understand the relationship between two variables through the simultaneous introduction of additional variables.

_____ 2. A variable used to divide the sample into subsets so that the original relationship can be examined in each of the subsets.

_____ 3. A variable that comes before both the independent and dependent variables.

_____ 4. A variable that comes between the independent and dependent variables.

_____ 5. Occurs when the partial relationships are essentially the same as original relationship.

_____ 6. Occurs when the original relationship can be explained in terms of an intervening variable.

_____ 7. Occurs when the original relationship can be explained in terms of an antecedent variable.

_____ 8. Occurs when partial relationships differ significantly from each other.

_____   9. A control variable that reverses the true relationship between the independent and dependent variables.

_____   10. The relationship between two variables when in a subset of cases defined by a third variable.

TRUE-FALSE QUESTIONS

T   F   1. Explanation is the term used to describe a spurious relationship where an original relationship is show to be false through the introduction of a test variable.

T   F   2. Replication is similar to explanation, except for the time placement of the test variable and the implications that follow from that difference.

T   F   3. The origins of the elaboration model go back to Likert and Guttman.

T   F   4. Partial relationships are compared to the zero-order relationship.

T   F   5. Specification occurs when the partial relationships differ significantly from each other.

T   F   6. A distorter variable conceals that relationship between an independent variable and a dependent variable.

T   F   7. The premise behind ex post facto hypothesizing is that all hypotheses must be subject to disconfirmation.

T   F   8. A test variable is different from a control variable.

T   F   9. Logically, elaboration analysis is difficult to do when the independent, dependent, and control variables have many categories.

T   F   10. The key difference between explanation and interpretation is whether the test variable comes before both the independent and dependent variables or between them.

REVIEW QUESTIONS

1. The basis of the elaboration model is *best* summarized as follows
    a. understanding the effects of antecedent variables.
    b. understanding the relationship between two variables through the controlled introduction of other variables.
    c. understanding the effects of intervening variables.
    d. explaining relationships between two variables that crop up inadvertently.
    e. supporting the findings of the partial relationships as compared with the original relationships.

2. The *key* aspect of the control variable that determines the type of elaboration is
    a. the number of categories.
    b. if there are more than one.
    c. the number of partials.
    d. the time order.
    e. the strength of the relationship.

3. Professor Kinkle conducted a survey in which he found that people over the age of 40 were more likely to put their parents in nursing homes than were people under age 40. When he controlled for age of parents, the results remained the same. This is an example of
    a. replication.
    b. explanation.
    c. specification.
    d. interpretation.
    e. ex post facto hypothesizing.

4. Researcher Lutter studied differences in men and women in degrees of stubbornness and found women to be more stubborn than men. When she controlled for zodiac signs, she found that the relationship held only under the sign of Taurus. This is an example of
    a. replication.
    b. explanation.
    c. specification.
    d. interpretation.
    e. ex post facto hypothesizing.

5.   The type of elaboration analysis that lends credibility and validity to a finding at the bivariate level is
   a.   replication.
   b.   specification.
   c.   explanation.
   d.   interpretation.
   e.   ex post facto hypothesizing.

6.   Examine the three tables below and then answer the question that follows.

A.   The original relationship between gender and voting:

	Male	Female
Voted	60%	50%
Did not vote	40%	50%
Total	100%	100%

B.   The relationship between gender and voting, but *only* for those **under** 35:

	Male	Female
Voted	20%	70%
Did not vote	80%	30%
Total	100%	100%

C.   The relationship between gender and voting, but *only* for those *over* 35:

	Male	Female
Voted	40%	50%
Did not vote	60%	50%
Total	100%	100%

Which type of elaboration is reflected in these tables?
   a.   Replication
   b.   Explanation
   c.   Specification
   d.   Interpretation
   e.   Reductionism

7.   In the tables in the preceding question, which is/are partial tables(s)?
       a.   A
       b.   B
       c.   C
       d.   all three
       e.   B and C only

8.   Professor Delphia found a relationship between gender and current occupational status. When he controlled for socioeconomic status background, the relationship disappeared. This is an example of
       a.   replication.
       b.   explanation.
       c.   specification.
       d.   interpretation.
       e.   ex post facto hypothesizing.

9.   In the above example socioeconomic status is a/an
       a.   independent variable.
       b.   antecedent variable.
       c.   intervening variable.
       d.   suppressor variable.
       e.   distorter variable.

10.   Professor Rodriquez found a strong positive relationship between participation in high school extracurricular activities and occupational success. She then controlled for social class background (when the respondents were very young) and the relationship vanished. Which type of elaboration is reflected in this example?
       a.   replication
       b.   specification
       c.   interpretation
       d.   explanation
       e.   ex post facto hypothesizing

11.   A spurious relationship is most likely to be identified with which type of elaboration?
   a.   replication
   b.   explanation
   c.   interpretation
   d.   specification
   e.   ex post facto hypothesizing

12.   Which of the following is *not* a limitation to the basic elaboration paradigm? Or are they all limitations?
   a.   It assumes an initial relationship.
   b.   It does not specify what constitutes a significant difference between the original and partial relationships.
   c.   It ignores the fact that partials may be stronger than the original relationship.
   d.   It focuses primarily on dichotomous variables.
   e.   All are limitations.

13.   Professor Luan is conducting research on the relationship between the number of hours a student studies and the grade received on a test. Inadvertently she discovers that the number of pizzas consumed the previous day is related to test performance. She concludes that students who take time out for themselves to relax are more likely to get higher grades than those who do not. This is an example of
   a.   replication.
   b.   explanation.
   c.   specification.
   d.   interpretation.
   e.   ex post facto hypothesizing.

14. Originally, when Professor Elizabeth examined the relationship between the number of mice in an apartment and the emotional well-being of the tenants, he found no relationship. Stubborn researcher that he is, he decided to control for the gender of the tenant. He found that for women, the more mice in the apartment the better their emotional well-being, although the opposite was true for men. In this example, gender is a/an
    a. independent variable.
    b. antecedent variable.
    c. intervening variable.
    d. suppressor variable.
    e. distorter variable.

15. Elaboration analysis was originally developed by
    a. Babbie and Glenn.
    b. Glenn and Stouffer.
    c. Stouffer and Lazarsfeld.
    d. Lazarsfeld and Miller.
    e. Miller and Durkheim.

16. In elaboration analysis, the original relationship that exists before introducing a control variable is known as the
    a. focal relationship.
    b. basis relationship.
    c. confounding relationship.
    d. suppressor relationship.
    e. zero-order relationship.

17. Dominic found that men were more likely to vote than women in his zero-order analysis. When he controlled for employment status, he found that women were more likely to vote in the partial for those unemployed. What type of variable is employment status in this instance?
    a. hidden
    b. cryptic
    c. suppressor variable
    d. attenuating variable
    e. distorter variable.

18. The hallmark of the elaboration paradigm is that it
    a. is a logical device for understanding more accurately the nuances of our data.
    b. can be used only with contingency tables.
    c. is vastly overrated as an analytical technique.
    d. is primarily a theory building activity.
    e. helped establish the prominence of sociology as a discipline.

19. Petras tested the relationship between number of siblings and current marital happiness, and controlled for highest education level. He discovered that the original relationship between number of siblings and current marital happiness disappeared within the partial tables for the various levels of educational attainment. What type of elaboration analysis does this example reflect?
    a. replicaiton
    b. explanation
    c. interpretation
    d. specification
    e. ex post facto hypothesizing

20. In elaboration analysis, the original relationship is recalculated for each of the categories of a control variable. The tables produced in this manner are called the
    a. zero-order relationships.
    b. elaborated relationships.
    c. multi variate relationships.
    d. partial relationships.
    e. distorter relationships.

DISCUSSION QUESTIONS

1. Briefly outline Stouffer's logical explanation for the anomalous finding that soldiers with more education were less resentful of being drafted than were soldiers with less education.

2. Without dealing with the four specific forms, summarize the logic of the elaboration model.

3. Compare and contrast the four specific outcomes of the elaboration model. Give an example of each that was not used in the text.

4. Explain the role of suppressor and distorter variables. Give an example of each that was not used in the text.

5. In what sense is ex post facto hypothesizing illegitimate? Why is it acceptable to formulate hypotheses after the fact in the special case of elaboration analysis?

EXERCISE 15.1

Name _____

Presented below are some hypothetical data describing white college graduates: (1) whether they are scored "high" or "low" on prejudice against minorities, (2) whether they attended college in the North or in the South, and (3) whether they were raised in the North or in the South. The final column in the table indicates the number of people having the set of characteristics shown to the left. For example, the first line of the table indicates that 20 people have "high" prejudice against minorities, went to college in the North, and grew up in the North; the second line indicates that 180 people have "low" minority prejudice, went to college in the North, and grew up in the North.

You are to use the elaboration model to test fully the following bivariate hypothesis about the impact of region of college on prejudice against minorities:

> Hypothesis: White college graduates who attended college in the South will be more prejudiced against minorities than those who attended college in the North.

In testing this hypothesis, complete the items that follow the data table.

PREJUDICE	REGION OF COLLEGE	REGION OF CHILDHOOD	N
High	North	North	20
Low	North	North	180
High	North	South	80
Low	North	South	20
High	South	North	05
Low	South	North	45
High	South	South	160
Low	South	South	40

(continued)

1. Identify the independent and dependent variables in the hypothesis. Construct the bivariate table that tests this basic hypothesis. Be sure you follow the guidelines for effective table presentation and analysis discussed in Chapter 15. Interpret the results.

2. Construct the other two bivariate tables that examine the relationship of the control variable first to the independent variable and then to the dependent variable of the original hypothesis. Interpret the results.

(continued)

3. Construct the trivariate table appropriate for assessing the effect of the control variable on the original relationship. Do this by presenting two versions of the original bivariate table, one for each attribute of the control variable. Analyze the results.

4. Which form of the elaboration model *best* represents the pattern of your results? Why?

EXERCISE 15.2

Name _____

This exercise resembles the preceding exercise. The variables in this case are: 1) the gender of automobile drivers, 2) whether they have driven "many" miles in their lives or "few" miles, and 3) whether they have had "many" accidents or "few" accidents.

You are to use the elaboration model to test fully the following hypothesis about the impact of gender on number of automobile accidents:

Hypothesis: Men have more automobile accidents than do women.

In testing this hypothesis, complete the items that follow the data table.

SEX	MILES DRIVEN	# OF ACCIDENTS	N
Women	Few	Many	20
Women	Few	Few	180
Women	Many	Many	80
Women	Many	Few	20
Men	Few	Many	05
Men	Few	Few	45
Men	Many	Many	160
Men	Many	Few	40

1. Identify the independent and dependent variables in the hypothesis. Construct the bivariate table that tests this basic hypothesis. Be sure you follow the guidelines for effective table presentation and analysis discussed in Chapter 15. Interpret the results.

(continued)

2. Construct the other two bivariate tables that examine the relationship of the control variable first to the independent variable and then to the dependent variable of the original hypothesis. Interpret the results.

3. Construct the trivariate table appropriate for assessing the effect of the control variable on the original relationship. Do this by presenting two versions of the original bivariate table, one for each attribute of the control variable. Analyze the results.

4. Which form of the elaboration model *best* represents the pattern of your results? Why?

EXERCISE 15.3

Name _____

You are to perform an elaboration analysis on one of the bivariate tables you produced and analyzed in Exercise 14.7. If you have not completed that exercise, select several independent and dependent variables from the General Social Survey codebook in Appendix 1, and produce the bivariate tables. It will be easier to interpret the results if each of the variables you select has only a few attributes. Select a dichotomous control variable—one with only two attributes. If you cannot easily identify a relevant dichotomous control variable, consider recoding another control variable into two categories. Your instructor may provide additional instructions for the particular version of SPSS or other data analysis program that you will use.

1. Present the bivariate table that represents the original relationship between the independent variable and the dependent variable. Be sure you follow the guidelines for effective table presentation and analysis discussed in Chapter 14. Analyze the results.

(continued)

2. Present the trivariate table appropriate for assessing the effect of the control variable on the original relationship. Do this by presenting two versions of the original bivariate table, one for each attribute of the control variable. Analyze the results.

3. Which form of the elaboration model *best* represents your results? Why?

EXERCISE 15.4

Name _____

You are to perform another elaboration analysis on one of the bivariate tables you produced and analyzed in Exercise 14.7. You may use the same original bivariate table you used in Exercise 15.3 if you wish. If you have not completed Exercise 14.7, select several independent and dependent variables from the General Social Survey codebook in Appendix 1, and produce the bivariate tables. It will be easier to interpret the results if each of the variables you select has only a few attributes. This time select a control variable with three or four attributes. Your instructor may provide additional instructions for the particular version of SPSS or other data analysis program that you will use.

1. Present the bivariate table that represents the original relationship between the independent variable and the dependent variable. Be sure you follow the guidelines for effective table presentation and analysis discussed in Chapter 14. Analyze the results.

(continued)

2. Present the trivariate table appropriate for assessing the effect of the control variable on the original relationship. Do this by presenting versions of the original bivariate table for each attribute of the control variable. Analyze the results.

3. Which form of the elaboration model *best* represents your results? Why?

4. Comment on differences in analysis procedures and difficulties between this exercise (with a nondichotomous control variable) and Exercise 15.3 (with a dichotomous control variable).

ADDITIONAL INTERNET EXERCISES

1. Read the example of elaboration analysis using General Social Survey data at http://www.csub.edu/ssric-trd/SPSS/SPSS11-8/11-8.htm (read up to "Regression Analysis"). What type of elaboration analysis is reflected in the example? Repeat the analysis using the 2006 GSS data accompanying this text.

2. Visit http://www.apsu.edu/oconnort/crimtheo2.htm, which "attempts to list all the possible relationships in a parsimonious, 3-variable model." Select three types not discussed in Babbie and give examples.

Chapter 16

Statistical Analyses

OBJECTIVES

1. Differentiate descriptive statistics from inferential statistics.

2. Define raw-data matrix.

3. Define data reduction.

4. Define measures of association and explain the logic behind such measures, using the proportionate reduction of error principle.

5. Describe the logic and calculation of lambda, gamma, and Pearson's *r*.

6. Describe the relationship between levels of measurement and measures of association.

7. Briefly describe regression analysis as an analytical technique.

8. Summarize and note the utility of each of the following types of regression analysis: linear regression, multiple regression, partial regression, and curvilinear regression.

9.   Summarize the cautions that should be applied when using regression analysis.

10.   Outline the logic, the calculation, and the interpretation of the standard error.

11.   Define and state the relationship between confidence level and confidence interval and give an example.

12.   List three assumptions underlying inferential statistics.

13.   Define tests of significance and link this concept to that of standard error.

14.   Show how the level of significance is interpreted.

15.   Using the null hypothesis, describe how chi square is calculated and interpreted.

16.   Describe the logic of the *t*-test.

17.   Summarize three dangers in interpreting the results of tests of significance.

18.   Define and note the utility of path analysis.

19.   Define and note the utility of time-series analysis.

20.   Define and note the utility of factor analysis.

21.   Define and note the utility of one-way analysis of variance and two-way analysis of variance.

22.   Define and note the utility of discriminant analysis.

23.   Define and note the utility of log-linear models.

24.   Define and note the utility of geographic information systems.

OUTLINE

1.   Descriptive statistics
     a,   Data reduction
     b.   Measures of association
     c.   Regression analysis

2.   Inferential statistics
     a.   Univariate inferences
     b.   Tests of statistical significance
     c.   The logic of statistical significance
     d.   Chi square
     e.   *t*-test

3.   Other multi variate techniques
     a.   Path analysis
     b.   Time-series analysis
     c.   Factor analysis
     d.   Analysis of variance
     e.   Discriminant analysis
     f.   Log-linear models
     g.   Geographic information systems (GIS)

SUMMARY

Social scientists employ two broad types of statistics. Descriptive statistics yield the capability for describing data in manageable form, and inferential statistics yield conclusions about a population from a study of a sample of that population.

Descriptive statistics serve a data reduction function. Although a raw-data matrix contains all the original information about the cases in a study, this format is very inefficient for presenting and analyzing data. Such univariate descriptive statistics as the mean, mode, and median efficiently summarize the central tendency in a frequency distribution. Measures of association perform a data reduction function by summarizing the joint frequency distributions of two variables.

Many measures of association are based on the proportionate reduction of error (PRE) principle, which incorporates a comparison between predicting a variable given its own distribution and given the joint distribution with another variable. The PRE principle is modified according to the level of measurement of the variables. Lambda is used with nominal variables, and its calculation involves using the mode. Its value ranges from zero (no association) to one (a perfect relationship). Gamma is used for ordinal variables and is based on comparing all possible pairs of cases for their relative rank ordering on both variables. The number of pairs of cases with the same ranking on the two variables is compared with the number of pairs of cases having the opposite ranking on the two variables. The values for gamma may vary from -1.0 to +1.0, thereby incorporating the direction as well as the strength of the association.

Pearson's r is used for interval and ratio level variables and uses the mean in applying the PRE principle. The use of a regression line minimizes the number of errors and yields the unexplained variation. The explained variation is the difference between the total variation and the unexplained variation. Regression analysis helps understand the Pearson's r. Regression analysis is a method by which the relationship between two or more variables can be specified in the form of a mathematical equation, known as the regression equation. This equation is a technique for representing the line that comes closest to the distribution of points in a joint frequency distribution. The general form of the equation in the case of one dependent and one independent variable is $Y = a + bX$, where X and Y represent the independent and dependent variables respectively; b represents the slope of the regression line; and a represents the Y intercept–the value of Y when X is equal to zero.

The bivariate linear regression model can be extended to the analysis of more than one independent variable, known as multiple regression analysis. This procedure allows a researcher to determine the relative contributions of each of several independent variables on a given dependent variable. Partial regression analysis resembles the elaboration model in that the equation summarizing the relationship between two variables is computed on the basis of a control variable. Curvilinear regression analysis enables the researcher to deal with relationships that fail to meet the linear model.

Caution should be applied in using regression analysis because the use of such analysis for statistical inference is based on the same assumptions made for correlational analysis: simple random sampling, the absence of nonsampling errors, and continuous interval data. Hence regression analysis can be used for interpolation (estimating cases lying between those observed) but is less useful for extrapolation (estimating cases that lie beyond the range of observations). Inferential statistics can be used to estimate the expected range of error involved in inferring a population value from a sample value. This statistic is known as the standard error and involves two components. The confidence level establishes how sure the researcher can be that a population value lies within a given range, the confidence interval. Several assumptions apply in inferential statistics: the sample must be drawn from the population about which inferences are being made, simple random sampling is assumed, and nonsampling errors are ignored.

Tests of statistical significance are used to specify the probability that an observed relationship between two variables (based on sample data) could be due only to sampling error. Statistical significance differs from substantive significance–whether a relationship is substantial enough to make a meaningful difference. A test of statistical significance such as chi square is actually a test of the null hypothesis that there is no relationship in the population from which the sample was selected. Chi square helps determine if the differences in the values on the dependent variable along categories of the independent variable are significantly different from the values we would find if there was no relationship between the independent and dependent variables. The level of significance indicates the probability of obtaining a sample with a degree of association at least as great as that in the observed sample data from a population in which there is, in fact, no relationship.

Chi square is used for testing the statistical significance of relationships found in nominal or ordinal data, while the t-test is used with interval or ratio level dependent variables by examining differences in group means. The value of the *t*-test will be larger, showing greater likelihood of a statistically significant difference, under several conditions: the size of the difference between the mean increases, the size of the sample increases, and when variations of values within each group is smaller.

Several dangers pertain to interpreting the results of statistical significance. First, there are no objective tests of substantive significance. Because low measures of association will be statistically significant in large samples, it is important to assess both statistical and substantive significance. Second, it is inappropriate to apply significance tests to data obtained from a total population since no sampling error occurs in such studies. Third, tests of significance are based on sampling assumptions that are frequently not met by the actual sampling design.

Other multivariate techniques can be applied to better understand the relationships among several variables. Path analysis is an extension of multiple regression analysis and is an efficient way to describe causal relationships among variables. It provides a graphic picture of both direct and indirect relationships among a series of independent variables as they influence a final dependent variable. This strategy requires a prior specification of the appropriate causal model by the researcher. Path coefficients are used to represent the strengths of the relationships of pairs of variables with the effects of all other variables in the model held constant.

Regression analysis may be used to analyze time-series data; such data represent changes in one or more variables over time. Such an analysis could express the long-term trend in a regression format and provide a way to test explanations for the trend. Time-lagged regression analysis enables even more sophisticated types of time-series analysis by linking changes in one variable at one point in time with changes in another variable at another point in time.

Factor analysis is somewhat different than regression analysis. It is used to discover underlying dimensions among the variations in values of many variables. Computer analysis is used to generate these dimensions, and the factor loadings provided indicate the correlation between each variable and each factor. As such, it is an excellent example of data reduction. However, factors are computer generated without any regard to substantive meaning, and the procedure will always generate some factors, making hypotheses useless.

Analysis of variance (ANOVA) employs the logic of statistical significance to examine differences on a dependent variable in terms of an independent variable. The independent variable is usually nominal and the dependent variable is usually interval or ratio. We compare the differences separating group means with the variations found within each group. The results are expressed in terms of statistical significance. One-way ANOVA involves one independent variable and one dependent variable, while two-way ANOVA involves more than two independent variables and one dependent variable.

The logic of discriminant analysis resembles that of multiple regression, but the dependent variable can be nominal. The purpose is similar to that of regression analysis: predicting values on the dependent variable given values on the independent variables. Log-linear models provide an approach for studying the impact of multiple non-dichotomous nominal level independent variables on a non-dichotomous nominal level dependent variable. This technique is based on specifying models that describe the interrelationships among variables and then comparing expected and observed table-cell frequencies. Multiple sets of relationships are included: between the dependent and independent variables, between pairs of independent variables, and three-variable (or more) relationships. Log-linear analysis helps identify which independent variables are the most salient for understanding the dependent variable.

Geographic information systems present aggregated data on geographic units in a graphical format. GIS enables more intuitive analyses of how social behaviors and characteristics vary across counties, states, or other geographical areas.

TERMS

1. chi square
2. confidence interval
3. confidence level
4. correlation matrix
5. curvilinear regression
6. data reduction
7. degrees of freedom
8. descriptive statistics
9. discriminant analysis
10. expected frequencies
11. explained variation
12. extrapolation
13. factor analysis
14. factor loadings
15. gamma
16. Geographic Information Systems
17. independence
18. inferential statistics
19. interpolation
20. joint distribution
21. lambda
22. level of significance
23. linear regression
24. log-linear models
25. measure of association
26. null hypothesis
27. one-way analysis of variance
28. path analysis
29. path coefficients
30. Pearson's *r*
31. proportionate reduction in error
32. raw-data matrix
33. regression analysis
34. regression equation
35. regression line
36. representativeness
37. residual
38. slope
39. standard error
40. substantive significance
41. *t*-test
42. tests of significance
43. time-lagged regression analysis
44. total variation
45. two-way analysis of variance

MATCHING

_____ 1. The type of statistics that describes data in a manageable form.

_____ 2. The graphic depiction of regression analysis.

_____ 3. The general type of statistic that indicates how strongly related two variables are.

_____ 4. The principle on which most measures of association are based.

_____ 5. An ordinal measure of association that uses pairs as the basis of comparison.

_____ 6. A causal time-ordered model for understanding relationships.

_____ 7. A test of significance that compares observed with expected frequencies.

_____ 8. A method used to discover patterns among the variations in values of several variables.

_____ 9. A method of analysis similar to multiple regression, except that the dependent variable can be nominal.

_____ 10. The correlations between each variable and each factor to show how much each variable is connected to each factor.

TRUE-FALSE QUESTIONS

T   F   1. The *t*-test is used with nominal independent and dependent variables.

T   F   2. Chi square is a measure of association.

T   F   3. Factor analysis is used to condense many items into fewer categories of a concept.

T   F   4. Path analysis is a time-ordered type of regression analysis.

T   F   5. Pearson's product-moment correlation is used with interval or ratio level variables.

T   F   6. Inferential statistics are based on the proportionate reduction of error principle.

T   F   7. Gamma is used with ordinal level variables.

T   F   8. Time-series analysis helps express the long-term trend in a regression format.

T   F   9. Discriminant analysis has a logic similar to regression analysis but uses nominal dependent variables.

T   F   10. Geographic Information Systems use aggregated data presented in graphical format.

REVIEW QUESTIONS

1.   The *major* purpose of descriptive statistics is
   a.   data reduction.
   b.   making inferences from the sample to the population.
   c.   establishing generalizability.
   d.   determining measures of association.
   e.   summarizing associations among variables.

2.   Professor Green wishes to employ inferential statistics. Which of the following is *not* an assumption he should keep in mind in using such statistics?
   a.   Samples must be drawn from the population to which he wishes to make inferences.
   b.   He should use simple random sampling, with replacement.
   c.   He should attain a 50 percent or better response rate.
   d.   These statistics ignore nonsampling errors.
   e.   All are correct assumptions.

3.   The proportionate reduction of error principle involves comparing

    a.   errors made with expected values with errors made with observed variables.

    b.   percentages across categories of the independent variable.

    c.   errors made at the bivariate level with errors made at the multivariate level.

    d.   errors made on the dependent variable alone with errors made knowing both the independent and dependent variables.

    e.   chi square with lambda.

4.   Professor Pegret has a hunch that Capricorns are usually Jewish. To assess the relationship between zodiac sign and religion, she should use which statistic?

    a.   lambda

    b.   gamma

    c.   chi square

    d.   Pearson's *r*

    e.   regression analysis

5.   The regression line drawn through a scattergram plotting cases on two variables

    a.   is usually not a straight line.

    b.   is drawn closest to the extreme values.

    c.   is drawn from left to right and from right to left.

    d.   can only be drawn if there are more than 50 cases.

    e.   is drawn as to minimize the deviations between each case and the line.

6.   Which of the following *best* represents what gamma measures?

    a.   similarities and dissimilarities in rank ordering of each pair of cases on one variable

    b.   a comparison of the rank order of all cases on one variable with the rank order on another variable

    c.   similarities and dissimilarities in rank ordering of each pair of cases across two variables

    d.   a comparison of the medians on two variables

    e.   the reduction in error attained by shifting from one variable to two

7. To find out the relationship between age (in years) and frequency of movie attendance, a researcher would use which of the following statistics?
    a. chi square
    b. regression analysis
    c. Pearson's *r*
    d. lambda
    e. gamma

8. Professor George has studied the relationship between age and church attendance in a sample of 50 people. He used regression analysis. The oldest person in his sample is 58. A friend wants him to estimate the church attendance for someone who is 64. Which strategy would be best?
    a. specification
    b. interpolation
    c. partial regression analysis
    d. extrapolation
    e. multiple regression analysis

9. The **key** factor determining which measure of association to use is
    a. the size of the sample.
    b. the level of measurement of the variables.
    c. how many variables are involved.
    d. the strength of the chi square.
    e. the strength of the relationship.

10. Professor Shoemaker wishes to develop an equation predicting this year's innovation rate in ten corporations based on last year's budget change and on the number of new people hired in the last two years. Which one of the following would be **best**?
    a. time-lagged regression analysis
    b. partial regression analysis
    c. quasi-experimental design
    d. time-series analysis
    e. factor analysis

11.   Professor Ford has developed an index of 50 items reflecting five dimensions of the concept *love*. Which technique should she use to determine if the empirical structure among the items represents her theoretical specification of the dimensions?
    a.   smallest-space analysis
    b.   factor analysis
    c.   multiple regression analysis
    d.   partial regression analysis
    e.   log-linear analysis

12.   Professor Geertsen calculated a standard error of two percent in his study of support for a tuition reduction at a university, and found that 40 percent supported the motion. What is the confidence interval for the 95 percent confidence level?
    a.   36 to 44
    b.   4
    c.   38 to 43
    d.   6
    e.   plus or minus 2

13.   Which of the following is an example of the null hypothesis?
    a.   There is no difference between males and females on voting.
    b.   There is a difference between males and females on voting.
    c.   Males tend to vote more often than females.
    d.   The relationship between sex and voting is unknown.
    e.   All are examples of the null hypothesis.

14. A student calculated a chi square significance level of .01 for a table relating gender to liberalism for those 18-25 years old. She calculated a chi square significance level of .25 for the same two variables for those 26-35 years old. Which of the following *best* summarizes the findings?
    a. She can be more sure of a real relationship between gender and liberalism among those 18-25 years old.
    b. She can be relatively sure that the relationship between gender and liberalism is strong for both age groups.
    c. The relationship between gender and liberalism is stronger for those 18-25 years old than for those 26-35 years old.
    d. The level of association between gender and liberalism is quite low for both age groups.
    e. None of the above.

15. Babbie's stance regarding statistical assumptions is that
    a. they should never be violated.
    b. it is OK to violate them if doing so helps understand the data.
    c. it is OK to violate them once you've learned a lot about statistics.
    d. it is OK to violate them on very rare occasions, but you should report that you did so.
    e. it is OK to violate them since the results would be the same anyway.

16. A strategy similar to multiple regression analysis except that the dependent variable can be nominal is known as
    a. one-way analysis of variance.
    b. two-way analysis of variance.
    c. discriminant analysis.
    d. Geographic Information Systems
    e. factor analysis.

17. Gamma is calculated by using which two quantities?
    a. the number of pairs available in a data set and the number of pairs having the same ranking on the two variables
    b. the number of pairs having the same ranking on two variables and the number of pairs having the opposite ranking on the same two variables
    c. the number of pairs having the opposite ranking on two variables and the number of pairs available in a data set
    d. the number of pairs available in a data set and the number of pairs we might expect to see if a relationship exists
    e. the number of people whose total score is an odd number and the number of people whose total score is an even number

18. For interval level variables you would minimize errors in guessing values on one variable, using the proportionate reduction in error principle, by guessing the
    a. standard deviation
    b. mode
    c. mean
    d. median
    e. interquartile range

19. Nimby was asked to create a scheme to highlight the measures of association among all the variables in his data set. What is such a scheme called?
    a. inverse matrix
    b. matching matrix
    c. gamma matrix
    d. the movie matrix
    e. correlation matrix

20.　Luke wants to be able to predict divorce given a couple's length of courtship. He wishes to control for age at marriage. That is, he wants to be able to infer a predicted divorce status for a particular length of courtship, controlling for age at marriage. Which one of the following approaches would be best one to use?

a.　regression analysis
b.　partial regression analysis
c.　factor analysis
d.　discriminant analysis
e.　smallest-space analysis

DISCUSSION QUESTIONS

1.　Compare and contrast the functions of descriptive and inferential statistics.

2.　Explain the proportionate reduction in error (PRE) principle as it relates to measures of association, and show how it applies in the case of lambda.

3. Explain how the concept of statistical significance is linked to sampling.

4. Describe what the various forms of regression analysis have in common. Then briefly differentiate among the following: linear regression, multiple regression, partial regression, curvilinear regression.

EXERCISE 16.1   Name _____

Presented below are hypothetical data representing the relationship between two nominal level variables, race and political party affiliation. Select a measure of association appropriate for examining the relationship and compute it. Please show your calculations. Interpret the results.

		Race			
		White	Black	Other	Total
	Republican	70	32	10	115
Party	Democrat	20	50	40	110
	Independent	10	15	50	75
	Total	100	100	100	300

1.   Which measure of association did you select? Why?

2.   What is the calculated value of this statistic? Show your calculations.

3.   Interpret this value.

EXERCISE 16.2

Name _____

Presented below are hypothetical data representing the relationship between two ordinal level variables, education and religiosity. Select a measure of association appropriate for examining the relationship and compute it. Please show your calculations.

		Education			
		Less than high school diploma	High school diploma	Some college or more	Total
Religiosity	Low	70	25	20	115
	Moderate	50	90	20	160
	High	20	30	60	110
	Total	140	145	100	385

1. Which measure of association did you select? Why?

2. What is the calculated value of this statistic? Show your calculations.

3. Interpret this value.

EXERCISE 16.3

Name _____

Calculate chi square for the data in the table in Exercise 16.1 (assume a random sample was used). Fill in the worksheet below. Cells are numbered across, not counting totals. Hence, white Republicans are cell 1, black Republicans are cell 2, and so on. A shortcut technique for calculating expected values is to multiply the column total for a cell times the row total for the same cell and divide by the total sample size.

Cell #	Observed frequency	Expected frequency	Ob - Ex	$(Ob - Ex)^2$	$\dfrac{(Ob - Ex)^2}{E}$
1					
2					
3					
4					
5					
6					
7					
8					
9					

1. What is the value of chi square?

2. How many degrees of freedom does this table have?

3. What is the level of significance for the chi square value?

4. Interpret both the chi square value and the level of significance.

EXERCISE 16.4

Name _____

Exercise 14.7 involved producing several bivariate tables using the General Social Survey. Use the tables from that exercise or from Exercise 15.3 or Exercise 15.4. If you have not completed any of those exercises, select several independent and dependent variables from the codebook in Appendix 1 of this workbook and produce the bivariate tables (be sure to request the appropriate statistics). Be sure to produce at least one table with both variables at the nominal level and at least one table with both variables at the ordinal level.

1.   Present your table with both variables at the nominal level.

2.   Interpret both chi square and lambda (or other statistics suggested by your instructor).

3.   Present your table with both variables at the ordinal level.

4.   Interpret both chi square and gamma (or other statistics suggested by your instructor).

EXERCISE 16.5

Name _____

In this exercise you will be working with the linear regression equation:

$$Y = 3 + 1.5X$$

Use the graph provided below.

```
   14

   13

   12

   11

   10

    9

    8

Y   7

    6

    5

    4

    3

    2

    1

    0
        0  1  2  3  4  5  6  7  8  9  10  11  12  13  14
                           X
```

(continued)

1. Compute the estimated values of Y for each of the following values of X.

    a. $X = 3, \ Y =$

    b. $X = 2.4, \ Y =$

    c. $X = 4.6, \ Y =$

2. Plot these estimated values of Y, using the designations A, B, C on the graph above.

3. Identify the Y intercept (or constant) and the slope for the above line.

ADDITIONAL INTERNET EXERCISES

1. Visit the federal social statistics briefing room at http://www.whitehouse.gov/http://www.whitehouse.gov/fsbr/ssbr.http://www.whitehouse.gov/fsbr/ssbr.html and click on a topic of interest. Show how some of the techniques and statistics noted in the chapter are reflected in your topic.

2. Read the *Special Report of the World's Women 2005. Progress in Statistics* at http://unstats.un.org/unsd/demographic/products/indwm/ww_specialRep.pdf. This report is issued by the United Nations. How does your understanding of Chapter 16 in the text better enable you to understand this report?

3. Review recent entries at the Social Science Statistics Blog, http://www.iq.harvard.edu/blog/sss/ and select three posts to review. Connect the content of these three posts to content in the chapter.

Chapter 17

Reading and Writing Social Research

OBJECTIVES

1. Provide advice for organizing a review of the literature.

2. Provide advice for reading journal articles.

3. Provide advice for reading research monographs.

4. Identify questions to ask when assessing theoretical orientations in research reports.

5. Identify questions to ask when assessing research designs in research reports.

6. Identify questions to ask when assessing measurement in research reports.

7. Identify questions to ask when assessing sampling in research reports.

8.   Identify questions to ask when assessing experiments in research reports.

9.   Identify questions to ask when assessing survey research in research reports.

10.   Identify questions to ask when assessing field research in research reports.

11.   Identify questions to ask when assessing the analysis of existing statistics in research reports.

12.   Identify questions to ask when assessing evaluation research in research reports.

13.   Identify questions to ask when assessing data analysis in research reports.

14.   Identify questions to ask when assessing the quality of reporting in research reports.

15.   Identify some relevant Web sites for learning about research methods.

16.   Explain how search engines can be used to search Web sites.

17.   Provide advice for assessing data found on Web sites.

18.   Provide advice for Web citations.

19.   Identify the functions of scientific reporting.

20.   Explain how the intended audience affects writing a research report.

21.   Differentiate the following types of research reports: research notes, working papers, professional papers, articles, and books.

22.   Compare the aims of research reports.

23.   Explain the importance of stating the purpose and providing an overview in research reports.

24.   Explain the role of reviewing the literature in research reports.

25.   Provide advice for avoiding plagiarism.

26.   Explain the role of describing the study design and execution in research reports.

27.   Explain the role of describing the analysis and interpretation in research reports.

28.   Explain the role of the summary and conclusions in research reports.

29.   Provide guidelines for reporting analyses in research reports.

30.   Identify some strategies for going public with research findings.

31.   Indicate relevant ethical considerations in reading and writing social research.

OUTLINE

1.   Organizing a review of the literature
       a.   Journals versus books
       b.   Evaluation of research reports

2.   Using the Internet wisely
       a.   Some useful Websites
       b.   Searching the Web
       c.   Evaluating the quality of Internet materials
       d.   Citing Internet materials

3.   Writing social research
       a.   Some basic considerations
       b.   Organization of the report
       c.   Guidelines for reporting analyses
       d.   Going public

4.   Ethics of reading and writing social research

SUMMARY

Important social scientific research and effective communication are tightly connected. It is often difficult to read and understand research reports, but it is important to learn how to communicate research results effectively. Most researchers begin with a review of the literature, and you should organize your search of the literature around the key concepts you wish to study. Search engines can help locate relevant studies. Sometimes it is useful to review the citations in studies you find to identify core references.

Several guidelines can be followed to maximize your reading of research articles. Begin with the abstract, which will tell you the purpose of the research, the methods used, and the major findings. The abstract serves two functions. First, it helps you decide if you want to read the rest of the article. Second the abstract provides a framework for reading the rest of the article. The summary or conclusions section will give you a more detailed view of the article. Then skim the article, paying particular attention to section headings and tables and graphs. When you read the entire article you will have a better grasp of the overall structure of the article. Skim the article quickly one more time after you have finished reading it. One way to see if you understand the article is to explain it to someone else.

Follow a similar approach when reading research monographs. These longer research reports address the same issues as do articles and follow a similar structure. The preface and opening chapter review the purpose of the study, the methods used, and the main findings. Follow a similar approach with each chapter by skimming the opening paragraphs and summary before reading the chapter. Depending on your reasons for reading the monograph, it is fine to skip or skim portions of the volume.

Examine the various dimensions of the research process when you evaluate research reports. Regarding theoretical orientations, identify the theoretical orientations and paradigms employed. Identify and assess hypotheses presented. Assess the connection between the theoretical orientations and the methods used. Regarding research design, identify the purpose of the study. Learn about who conducted the research and why. Identify the units of analysis and determine if the design is cross-sectional or longitudinal. Be sure the design fits the purposes of the study. Regarding measurement, identify the concepts used and the measures developed for each concept. Establish reliability and validity for the measures. Assess the appropriateness and quality of composite measures.

Regarding sampling, decide if sampling was appropriate or if it would have been better to study the entire population. Determine the sampling design used and its appropriateness. Identify the population to which generalizations are to be made. Scrutinize the sampling frame to make sure it adequately represents the population of interest. Determine if probability or nonprobability designs were used correctly. Determine if the sample size was adequate and if there is a large proportion of nonrespondents. See if the researcher has tested for representativeness. Decide if the conclusions made appropriately apply to the population studied.

Regarding experiments, identify the dependent variable–the outcome–and the independent variable–the stimulus. Determine if other relevant variables have been considered and measured appropriately. Assess the measures for the outcome and stimulus variables. Determine if control groups were used appropriately. Assess possible factors that might influence internal validity. See if external validity was addressed. Regarding survey research, look over the instrument to see if the questions follow the guidelines discussed in the survey research chapter. For example, make sure the responses are mutually exclusive and exhaustive, make sure the questions are clear and relevant, make sure the questions are not double barreled, and make sure that no negative items were used. If secondary analysis was employed, assess the quality and relevance of the data set used.

Regarding field research, determine which theoretical paradigms inform the research. See if the goal was to test hypotheses or to generate theory. Identify the main variables and their measures. Assess reliability and validity. See if the conclusions drawn match the data. Determine how generalizable the results are and if the researcher makes excessive claims of generalizability. Explain the role of the researcher and determine the appropriateness of that role. Assess any ethical difficulties. See if the researcher's own identity or background may have affected the research process or the results. Regarding content analysis, identify the key variables and the source and form of the data. Determine if the appropriate unit of analysis was employed and if appropriate analytical techniques are used in quantitative and qualitative analyses. Regarding the analysis of existing statistics, find out who originally collected the data and how they were collected. Identify the unit of analysis and its appropriateness. See if the operational definitions developed by the original researcher adequately fit the purposes of the current researcher. Regarding comparative and historical research, determine the purpose of the study and the unit of analysis. Identify the key variables and decide if the range of data used is appropriate to the analysis. Uncover the sources used for historical documents so that you can assess their potential bias.

Regarding evaluation research, identify the social intervention and how it was measured. Assess reliability and validity. See if appropriate people were observed. Look carefully at how "success" was defined and examine on what basis the researcher concluded that the intervention was a success or failure. Examine who sponsored the research and who did the research to assess the objectivity of the study. Regarding data analysis, see if quantitative or qualitative analyses were done appropriately, given the purpose and design of the study. Examine how nonstandardized data have been coded. See if the researcher performed all relevant analyses. Be alert to spurious relationships. Make sure that the findings reported really matter and have not been overstated. Make sure that the conclusions fit the data. See if new patterns of relationships were observed that might provide the basis for grounded theories. Make sure the statistical techniques are used appropriately and that tests of statistical significance are used and interpreted correctly. Regarding reporting, see if the researcher placed the study in the context of previous research. Make sure the details of the study are reported fully and that shortcomings are identified.

The World Wide Web is now a very valuable resource for researchers. Many relevant sites exist, of varying degrees of quality. Particularly useful sites are those by the General Social Survey, the Bureau of the Census, and various governmental agencies and universities. You can use various search engines to research your topics. You may have to refine your search as you narrow the relevant sites. Finding data on the Web is easy; assessing its quality is more difficult. Several items deserve attention when you assess Web sites. Determine the author of the site to help you assess any biases and sloppiness in data presentation and manipulation. See if the site advocates a particular point of view and then assess the data presented accordingly. See if the site provides accurate and complete references, and if the data presented are current. Give more credibility to official data, such as that presented by governmental sites. University sites are also usually more credible. Finally, see if the data presented are consistent with data from other sites. When citing Web sites, be sure to include the URL, the date and time the site was accessed, author and title, and cite page numbers when a site references a print version.

Writing about social research requires effort to do so well. You may wish to review books on effective writing. Understand that scientific reporting has several functions. First, your report should communicate a body of specific data and ideas. Second, you should see your report as a contribution to the general body of scientific knowledge. Finally, your report should stimulate and direct further study.

Some basic considerations apply in writing about social research. Tailor your report to your audience. Use a research note for a short report to be published in a journal. When preparing a report for a sponsor, focus on issues of concern to the sponsor. Use a working paper to share a tentative version of your report so that you can solicit feedback from peers. Professional papers are given at professional meetings and should be brief and direct. Articles published in journals are the most popular type of research report and run about 25 typed pages. A book is the most prestigious form of research report. Research reports address the different purposes of research: exploration, description, explanation, hypothesis testing, and application.

A general format exists for presenting research data. Begin with the purpose and overview of your study. Next review the relevant literature and show how your study fits in. Be sure to avoid plagiarism when doing so. You must give a complete citation when using another author's exact words. You must avoid paraphrasing another's work and presenting the results as your own work. You should not even present another's ideas as your own, even if you use different words to express those ideas. The study design and execution come after the literature review, followed by the analysis and interpretation of the results. The presentation of data, the manipulation of those data, and your interpretations should be integrated into a logical whole. Finish your report with the summary and conclusions.

There are additional guidelines for reporting analyses. Be sure to present as much detail as possible without clutter. Present quantitative data so that the reader can recompute them. Provide enough detail so that a researcher could replicate your results using secondary analysis. If reporting on qualitative research, provide enough detail so that the reader has a sense of having made the observations with you. Present all the data, including that which did not support your interpretations. Integrate supporting materials, such as tables and charts. Draw explicit conclusions. Point to any qualifications or conditions warranted in the evaluation of the conclusions.

Students often go public with their research studies by presenting at the annual meetings of professional associations. These presentations typically are about 15-20 minutes long. To pursue publication, find journals that publish articles on the topic of your research. Finally, ethical issues enter the reading and writing of social research. Be careful to avoid reviewing only the literature that supports your point of view.

TERMS

1. abstract
2. article
3. book
4. plagiarism
5. professional papers
6. research monograph
7. research note
8. search engines
9. URL
10. working papers

MATCHING

_____ 1. A book-length research report.

_____ 2. A mechanism for identifying relevant Web sites for a given topic.

_____ 3. The acronym representing a Web address.

_____ 4. Found at the beginning of an article, this element tells the purpose of the
research, the methods used, and the major findings.

_____ 5. A brief summary of research found in journals.

_____ 6. A tentative presentation of research findings with an implicit request for
comments.

_____ 7. Presentations given at professional meetings.

_____ 8. Theft of another's words and/or ideas.

_____ 9. A full-length research report in a journal.

_____ 10. Also called a research monograph.

TRUE-FALSE QUESTIONS

T   F   1.   An abstract generally appears at the end of an article.

T   F   2.   Google is a popular type of abstracting system for research articles.

T   F   3.   It's a good idea when assessing Web sites to determine if a site is advocating a particular point of view.

T   F   4.   Generally, official data reported at government Web sites are more objective than materials at many other Web sites.

T   F   5.   A Web address is known as an XML.

T   F   6.   If you wish to publish a brief report on your research in a journal, you should consider the research note.

T   F   7.   Plagiarism is of particular concern in the purpose and overview section of the research report.

T   F   8.   It is generally a good idea to avoid drawing explicit conclusions in a research report so that the reader can form her or his own conclusions.

T   F   9.   It is generally best to organize a search of the literature around key concepts.

T   F   10.   One particular ethical concern in reading and writing social research is the risk of reviewing the literature with a special eye toward reports that support a point of view you favor.

REVIEW QUESTIONS

1.   The first thing you should read in a journal article is the
   a.   conclusions.
   b.   methods.
   c.   sampling methods.
   d.   theoretical background.
   e.   abstract.

2.   The abstract for a journal article serves which two functions?
   a.   establish the author's credibility and justify the methods
   b.   justify the methods and hint at the study's design
   c.   hint at the study's design and help you decide to read the article or not
   d.   help you decide to read the article or not and establish a framework for reading the article
   e.   establish a framework for reading the article and establish the author's credibility.

3.   The strategy to use to fully grasp what you've read is to
   a.   write your own abstract for the article.
   b.   contact the author and share your observations about the article.
   c.   find someone else to explain it to.
   d.   recalculate the statistics in the article.
   e.   draw out the policy implications of study.

4.   The best strategy for reading research monographs is to
   a.   read it from cover to cover.
   b.   follow the same approach as reading research articles.
   c.   skip all the methods discussions.
   d.   focus on the theory and policy.
   e.   look at all the pictures and tables.

5.   When you identify the hypotheses tested in a study, which aspect of research reports are you assessing?
   a.   theoretical orientations
   b.   research design
   c.   measurement
   d.   sampling
   e.   experiments

6.   When you identify the units of analysis and whether the study was longitudinal or cross-sectional, which aspect of research reports are you assessing?
   a.   theoretical orientations
   b.   research design
   c.   measurement
   d.   sampling
   e.   experiments

7. When you determine if indexes and scales have been used appropriately, which aspect of research reports are you assessing?
   a. field research
   b. sampling
   c. data analysis
   d. evaluation research
   e. measurement

8. When you determine if control groups have been used appropriately, which aspect of research reports are you assessing?
   a. survey research
   b. experiments
   c. field research
   d. measurement
   e. experiments

9. When you determine if the author has shared any flaws shortcomings in the study, which aspect of research reports you assessing?
   a. reporting
   b. sampling
   c. theoretical orientations
   d. survey research
   e. measurement

10. What is the most straightforward way to find information on World Wide Web?
    a. visit the *Statistical Abstracts* site
    b. visit the Web site for the textbook
    c. use only governmental sites
    d. use a search engine
    e. use Web addresses that end with .edu

11. The two biggest risks you face in getting information from the Web are
    a. there is too much information and it is disorganized.
    b. information is disorganized and many sites are biased.
    c. many sites are biased and sloppy.
    d. sloppiness and too much unethical research is reported.
    e. too much unethical research is reported and there is too much information.

12. Two particularly useful and unbiased types of Web sites are
    a. sites by governmental agencies and universities.
    b. sites by universities and interest groups.
    c. sites by interest groups and sites with any type of data.
    d. sites with any type of data and sites by famous authors.
    e. sites by famous authors and governmental agencies.

13. Which one of the following is *not* a function of scientific reporting? Or are they all functions?
    a. communicate a body of specific data and ideas
    b. a contribution to the general body of scientific knowledge
    c. stimulate and direct further inquiry
    d. promote the author's reputation
    e. all are functions

14. You want to prepare a research report noting your tentative observations, and you want to solicit comments from readers. Best to use would be a/an
    a. professional paper.
    b. working paper.
    c. book.
    d. research note.
    e. article.

15. When reporting on the works of others, such as in a liter review, it is particularly important to
    a. cover a wide representation of authors.
    b. cover a wide range of years.
    c. avoid plagiarism.
    d. consider your audience.
    e. use only sociological works.

16. In most cases, you should organize your search of the literature around
    a. the time order of the references.
    b. the importance of the researchers who wrote the articles.
    c. the nature of the sampling methods.
    d. the key concepts you wish to study.
    e. how much each reference supports your hypotheses.

17. Most researchers begin the design of a research project with
    a. a review of the literature.
    b. a description of sampling methods.
    c. securing funding.
    d. a preview of their findings.
    e. deciding if the study should be quantitative or qualitative.

18. When doing a literature review, which of the following is particularly useful?
    a. using the card catalog
    b. browsing the stacks where the books are stored
    c. using search engines
    d. talking to a librarian
    e. using the Dewey Decimal System

19. When students go public with their research studies, they typically use
    a. news briefs and TV appearances.
    b. TV appearances and blogs.
    c. blogs and professional papers.
    d. professional papers and published articles.
    e. published articles and news briefs.

20. Oral presentations at professional meetings typically last about
    a. 5-10 minutes.
    b. 10-15 minutes.
    c. 15-20 minutes.
    d. 20-25 minutes.
    e. 25-30 minutes.

DISCUSSION QUESTIONS

1. Outline Babbie's advice for reading journal articles and books.

2. Summarize Babbie's advice for evaluating the following aspects of research reports: theoretical orientations, research design, measurement, sampling, experiments, survey research, field research, content analysis, analyzing existing statistics, comparative and historical research, evaluation research, data analysis, and reporting.

3. Summarize Babbie's advice for evaluating the quality of Internet materials.

4. Explain what constitutes plagiarism and what does not constitute plagiarism. Give examples.

EXERCISE 17.1

Name _____

Locate an online social science research report by using your library's resources. Assess the article by using the criteria noted in the text, organized into the sections used in the text. Some sections may not be relevant for your particular article. Report the full bibliographic information for your source.

1.   Bibliographic citation:

2.   Theoretical orientations:

3.   Research design:

4.   Measurement:

5.   Sampling:

6.   Experiments:

7.   Survey research:

(continued)

8.   Field research:

9.   Content analysis:

10.   Analyzing existing statistics:

11.   Comparative or historical research:

12.   Evaluation research:

13.   Data analysis:

14.   Reporting:

EXERCISE 17.2

Name _____

Use a search engine to find three Web sites on abortion or sex education. Select sites that contain data. Cite the Web sites using the guidelines noted in the text. Assess each site according to the guidelines suggested by Babbie.

    1.   First Web site citation:

    2.   Assessment of first Web site.
        a.   Who is the author?

        b.   Does the site advocate a particular viewpoint?

        c.   Does the site give accurate and complete references?

        d.   Are the data up-to-date?

        e.   Are the data official?

        f.   Is it a university research site?

        g.   Do the data seem consistent with data from other sites?

(continued)

3. Second Web site citation:

4. Assessment of second Web site.
   a. Who is the author?

   b. Does the site advocate a particular viewpoint?

   c. Does the site give accurate and complete references?

   d. Are the data up-to-date?

   e. Are the data official?

   f. Is it a university research site?

   g. Do the data seem consistent with data from other sites?

5. Third Web site citation:

(continued)

6.   Assessment of third Web site.
       a.   Who is the author?

       b.   Does the site advocate a particular viewpoint?

       c.   Does the site give accurate and complete references?

       d.   Are the data up-to-date?

       e.   Are the data official?

       f.   Is it a university research site?

       g.   Do the data seem consistent with data from other sites?

EXERCISE 17.3

Name _____

Use your library's online journals to find three sources on a similar topic. Provide full bibliographic information on your sources. Write two versions of a summary of these sources, one that reflects plagiarism and one that does not. Explain the difference.

1. Bibliographic citations for the three articles:

2. Summary involving plagiarism:

3. Summary not involving plagiarism:

4. Explain what makes the first summary plagiarized and the second not plagiarized.

ADDITIONAL INTERNET EXERCISES

1. Review the search engine tutorial at http://www.pandia.com/goalgetter/ http://searchenginewatch.com/.to learn more about search engines and how to use them. Indicate what was most useful.

2. Access http://www.tilburguniversity.nl/services/library/instruction/www/onlinecourse/. This site provides a tutorial for learning about the Internet, the World Wide Web, the types of information available on the Internet, and how to search the Internet. Indicate the three most useful things you learned at this site.

3. Select a study to review at http://www.sirc.org/. Apply what you learned in Chapter 17 to the study.

Appendix 1

General Social Survey

The General Social Survey (GSS) is a survey done by the National Opinion Research Center on the attitudes, behaviors, and background characteristics of Americans. These surveys have occurred annually or biannually since 1972. Samples are full probability samples, and the samples reflect multistage area probability samples to the block level. At the block level, however, quota sampling is used with quota based on sex, age, and employment status. The primary sampling units employed are Standard Metropolitan Statistical Areas or nonmetropolitan counties, stratified by region, age, and race before selection. The units of selection at the second stage are block groups and enumeration districts, stratified according to race and income before selection. The third stage of selection involved blocks, which were selected with probabilities proportional to size. The final sampling units involved housing units and individuals.

The original sample sizes for the 1996 and 2006 studies were 4,559 and 9,535. After accounting for such things as vacant buildings, language problems, new dwelling units, and other factors at the sampling level and for refusals, break offs before completion, no one home, and other factors at the individual level, there were 3,814 and 4,510 completed cases, resulting in response rates of 76.1 and 71.2 percent. The "split-ballot" technique was used, where some questions were asked of part of the sample and other questions were asked of another part of the sample. Basic questions were asked of all participants. As a result, the totals for some of the variables you analyze may not total the numbers of completed cases noted above.

The data files you will be using will vary according to the version of SPSS (or other statistical program) that you will use. *SPSS Student Version* is limited to 50 variables and 1,500 cases, but the regular version has no limitations. If you will be using Student SPSS, you will be using either a combined 1996-2006 data file containing 750 cases from each year or separate data files of 1,500 cases for each year. In both situations, you will have access only to the 50 variables in the codebook with asterisks. The subset data files are random samples taken from the full data files.

You may wish to limit your analyses to only one year, do separate analyses for the two years, or combine both years in one analysis. Unless you specifically select a particular year, your analyses will include data from both years (if you are using the combined years file). Be very clear on which subgroups you will use in your analysis. If you use the combined sample of cases across both years to study attitudes toward abortion, for example, changes that occurred in attitudes between the years will not be evident and may even mask relationships you wish to examine between, say, gender and attitudes toward abortion. That is, if women were more likely support abortion in one year and men in the other year, a table relating gender to abortion support for the combined sample would show a negligible gender difference. All the variables in the file were asked in both 1996 and 2006.

Be sure you distinguish clearly between independent and dependent variables in your analyses. Many of the typically independent variables, such as gender, age, and race, are located toward the beginning of the data file. Some variables, such as how much people watch TV or their level of education, could be either independent or dependent, depending on your research question. Select an appropriate number of what you consider independent and dependent variables, or as many of each type as indicated by your instructor.

Once again, an asterisk indicates variables which *are* available in *SPSS Student Version*. *All* the variables are available in the regular version of SPSS.

* = available in *SPSS Student Version*.

YEAR*   1.   Year of sample.
                                     1996 or 2006

REGION*   2.   Region of interview.
                             1. New England
                             2. Middle Atlantic
                             3. East North Central
                             4. West North Central
                             5. South Atlantic
                             6. East South Central
                             7. West South Central
                             8. Mountain
                             9. Pacific

RACE*   3.   What race do you consider yourself?
1. White
2. Black
3. Other

SEX*   4.   Sex. (Coded by interviewer)
1. Male
2. Female

AGE*   5.   Age. (Determined by asking date of birth, actual age recorded.) Values reflect actual age.

AGE3*   6.   AGE variable recoded into thirds.
1. 18-33
2. 34-51
3. 52-89

MARITAL*   7.   Are you currently married, widowed, divorced, separated, or have you never been married? (Divorced and separated combined into one category.)
1. Married
2. Widowed
3. Divorced/separated
4. Never married

DEGREE*   8.   Highest degree.
0. Less than high school
1. High school
2. Associate/junior college
3. Bachelor's
4. Graduate

REALINC*   9.   Family income in constant dollars.

INCOME3*   10.   REALINC, family income in constant dollars, recoded into thirds.
1. Low
2. Moderate
3. High

FINRELA 11. Compared with American families in general, would you say your family income is far below average, below average, above average, or far above average? (Recoded into three categories.)
    1. Below average
    2. Average
    3. Above average

FINALTER 12. During the last few years, has your financial situation been getting better, worse, or has it stayed the same?
    1. Better
    2. Worse
    3. Stayed the same

SATFIN 13. We are interested in how people are getting along financially these days. So far as you and your family are concerned, would you say that you are pretty well satisfied with your present financial situation, more or less satisfied, or not satisfied at all?
    1. Pretty well satisfied
    2. More or less satisfied
    3. Not satisfied at all

SEI* 14. Hodge/Siegel/Rossi prestige scale score for respondent's occupation. (Actual score recorded.)

PRESTIG3* 15. SEI, respondent's occupational prestige, recoded into thirds.
    1. Low
    2. Moderate
    3. High

CLASS 16. If you were asked to name one of four names in your social class, which would you say you belong in: the lower class, the working class, the middle class, or the upper class?
    1. Lower class
    2. Working class
    3. Middle class
    4. Upper class

RELIG*

17. What is your religious preference? Is it Protestant, Catholic, Jewish, some other religion, or no religion?
    1. Protestant
    2. Catholic
    3. Jewish
    4. None
    5. Other

ATTEND*

18. How often do you attend religious services? (Recoded into four categories)
    1. Less than once a year
    2. Once a year through several times a year
    3. About once a month through almost weekly
    4. Every week or more

FUND

19. Fundamentalism/liberalism of respondent's religion.
    1. Fundamentalist
    2. Moderate
    3. Liberal

CHILDS*

20. How many children have you ever had? Please count all that were born alive at any time (including any you had from a previous marriage).
    0-7. Actual number (e.g., 3 means 3 children)
    8. Eight or more

CHILDS3

21. CHILDS, number of children, recoded into thirds
    1. None
    2. One or two
    3. Three through eight or more

SIBS*

22. How many brothers and sisters did you have? Please count those born alive, but no longer living, as well as those alive now. Also include stepbrothers and stepsisters, and children adopted by your parents.
    Values reflect actual number of siblings.

SIBS3

23. SIBS, number of siblings, recoded into thirds
    1. 0-2
    2. 3-4
    3. 5 or more

PARTYID*

24. Generally speaking, do you usually think of yourself as a Republican, Democrat, Independent, or what? (Recoded into four categories.)
    1. Democrat
    2. Independent
    3. Republican
    4. Other party

POLVIEWS*

25. I'm going to show you a seven-point scale on which the political views that people might hold are arranged from extremely liberal—point 1—to extremely conservative—point 7. Where would you place yourself on this scale? (Recoded into three categories.)
    1. Liberal
    2. Moderate
    3. Conservative

ZODIAC

26. Astrological sign of respondent.
    1. Aries   7. Libra
    2. Taurus   8. Scorpio
    3. Gemini   9. Sagittarius
    4. Cancer   10. Capricorn
    5. Leo   11. Aquarius
    6. Virgo   12. Pisces

TVHOURS*

27. On the average day, about how many hours do you personally watch television?
    (Actual number of hours recorded.)
    Actual number of hours

TVHOURS3

28. TVHOURS, number of hours watching television per day, recoded into thirds.
    1. 0 through 1 hours
    2. 2 through 3 hours
    3. 4 through 24 hours

NEWS

29. How often do you read the newspaper—every day, a few times a week, once a week, less than once a week, or never? (Last two categories combined.)
    1. Every day
    2. A few times a week
    3. Once a week
    4. Less than once a week
    5. Never

OWNGUN*

30. Do you happen to have in your home any guns or revolvers?
    1. Yes
    2. No

MAWRKGRW

31. Did your mother every work for pay for as long as a year, while you were growing up?
    1. Yes
    2. No

HEALTH*

32. Would you say your health, in general, is excellent, good, fair, or poor?
    1. Excellent
    2. Good
    3. Fair
    4. Poor

Please tell me whether or not *you* think it should be possible for a pregnant woman to obtain a *legal* abortion if. . .(applies to items 33-38)

ABANY*

33. The woman wants it for any reason?
    1. Yes
    2. No

ABDEFECT*   34. There is a strong chance of serious defect in the baby?
　　　　　　　　　　　1. Yes
　　　　　　　　　　　2. No

ABHLTH*   35. The woman's own health is seriously endangered by the pregnancy?
　　　　　　　　　　　1. Yes
　　　　　　　　　　　2. No

ABNOMORE*   36. She is married and does not want any more children?
　　　　　　　　　　　1. Yes
　　　　　　　　　　　2. No

ABRAPE*   37. She became pregnant as a result of rape?
　　　　　　　　　　　1. Yes
　　　　　　　　　　　2. No

ABSINGLE*   38. She is not married and does not want to marry the man?
　　　　　　　　　　　1. Yes
　　　　　　　　　　　2. No

AGED*   39. As you know, many older people share a home with their grown children. Do you think this is generally a good idea or a bad idea?
　　　　　　　　　　　1. A good idea
　　　　　　　　　　　2. Bad idea
　　　　　　　　　　　3. Depends (volunteered)

I am going to name some institutions in this country. As far as the *people running* these institutions are concerned, would you say you have a great deal of confidence, only some confidence, or hardly any confidence at all in them? (Applies to items 40-45.)

CONARMY   40. Military?
　　　　　　　　　　　1. A great deal
　　　　　　　　　　　2. Only some
　　　　　　　　　　　3. Hardly any

CONBUS   41. Major companies?
                    1. A great deal
                    2. Only some
                    3. Hardly any

CONEDUC*   42. Education?
                    1. A great deal
                    2. Only some
                    3. Hardly any

CONFED   43. Executive branch of the federal government?
                    1. A great deal
                    2. Only some
                    3. Hardly any

CONLEGIS   44. Congress?
                    1. A great deal
                    2. Only some
                    3. Hardly any

CONPRESS*   45. Press?
                    1. A great deal
                    2. Only some
                    3. Hardly any

GRASS*   46. Do you think the use of marijuana should be made legal
                    or not?
                    1. Should be made legal
                    2. Should not be made legal

DIVLAW*   47. Should divorce in this country be easier or more difficult
                    to obtain than it is now?
                    1. Easier
                    2. More difficult
                    3. Stay as is (volunteered)

POLHITOK*   48.   Are there any situations you can imagine in which you would approve of a policeman striking an adult male citizen?
1. Yes
2. No

GUNLAW*   49.   Would you favor or oppose a law which would require a person to obtain a police permit before he or she could buy a gun?
1. Favor
2. Oppose

SOCFREND*   50.   How often do you spend a social evening with friends who live outside the neighborhood? (Collapsed from 7 categories.)
1. Several times a week or more
2. Several times a month
3. Once a month
4. Several times a year
5. Once a year or less

SOCOMMUN*   51.   How often do you spend a social evening with someone who lives in your neighborhood? (Collapsed from 7 categories.)
1. Several times a week or more
2. Several times a month
3. Once a month
4. Several times a year
5. Once a year or less

SOCREL   52.   How often do you spend a social evening with relatives? (Collapsed from 7 categories.)
1. Several times a week or more
2. Several times a month
3. Once a month
4. Several times a year
5. Once a year or less

HAPPY* 53. Taken all together, how would you say things are these days—would you say that you are very happy, pretty happy, or not too happy?
   1. Very happy
   2. Pretty happy
   3. Not too happy

SATJOB 54. On the whole, how satisfied are you with the work that you do—would you say that you are very satisfied, moderately satisfied, a little dissatisfied, or very dissatisfied? (Responses recoded into three categories.)
   1. Very satisfied
   2. Moderately satisfied
   3. Dissatisfied

SEXEDUC* 55. Would you be for or against sex education in the public schools?
   1. For
   2. Against

TRUST 56. Generally speaking, would you say that most people can be trusted or that you can't be too careful in life?
   1. Most people can be trusted
   2. Can't be too careful
   3. Depends (volunteered)

HELPFUL* 57. Would you say that most of the time people try to be helpful, or that they are mostly just looking out for themselves?
   1. Try to be helpful
   2. Just look out for themselves
   3. Depends (volunteered)

GETAHEAD 58. Some people say that people get ahead by their own hard work; others say that lucky breaks or help form other people are more important. Which do you think is most important?
   1. Hard work
   2. Hard work and luck equally important
   3. Luck

FEAR   59. Is there any area right around here—that is, within a mile—where you would be afraid to walk alone at night?
                1. Yes
                2. No

POSTLIFE   60. Do you believe there is a life after death?
                1. Yes
                2. No

PREMARSX*   61. There's been a lot of discussion about the way morals and attitudes about sex are changing in this country. If a man and a woman have sex relations before marriage, do you think it is always wrong, almost always wrong, wrongly only sometimes, or not wrong at all?
                1. Always wrong
                2. Almost always wrong
                3. Wrong only sometimes
                4. Not wrong at all

HOMOSEX*   62. What about sexual relations between two adults of the same sex—do you think it is always wrong, almost always wrong, wrong only sometimes, or not wrong at all?
                1. Always wrong
                2. Almost always wrong
                3. Wrong only sometimes
                4. Not wrong at all

XMARSEX*   63. What is your opinion about a married person having sexual relations with someone other than the marriage partner—is it always wrong, almost always wrong, wrong only sometimes, or not wrong at all?
                1. Always wrong
                2. Almost always wrong
                3. Wrong only sometimes
                4. Not wrong at all

SEXFREQ*   64. About how often did you have sex in the last twelve months?

    0. Not at all
    1. Once or twice
    2. Once a month
    3. Two or three times a month
    4. Weekly
    5. Two or three times a week
    6. Four or more times a week

PARTNRS5   65. Including the past 12 months, how many sex partners have you had in the past five years?

    0. No partners
    1. One partner
    2. Two partners
    3. Three or four partners
    4. Five or more partners

XMOVIE   66. Have you seen an X-rated movie in the last year?

    1. Yes
    2. No

We are faced with many problems in this country, none of which can be solved easily or inexpensively. I'm going to name some of these problems, and for each one I'd like you to tell me whether you think we're spending too much money on it, too little money, or about the right amount. Are we spending too much, too little, or about the right amount on. . .(Applies to items 67-72.)

NATAID   67. Foreign aid?

    1. Too little
    2. About right
    3. Too much

NATCRIME*   68. Halting the rising crime rate?

    1. Too little
    2. About right
    3. Too much

NATARMS   69. The military, armaments, and defense?
                1. Too little
                2. About right
                3. Too much

NATEDUC*   70. Improving the nation's education system?
                1. Too little
                2. About right
                3. Too much

NATFARE   71. Welfare?
                1. Too little
                2. About right
                3. Too much

NATRACE   72. Improving the condition of Blacks?
                1. Too little
                2. About right
                3. Too much

FECHLD*   73. Please tell me whether you strongly agree, agree, disagree, or strongly disagree with this statement: a working mother can establish just as warm and secure a relationship with her children as a mother who does not work.
                1. Strongly agree
                2. Agree
                3. Disagree
                4. Strongly disagree

FEFAM*   74. Please tell me whether you strongly agree, agree, disagree, or strongly disagree with this statement: it is much better for everyone involved if the man is the achiever outside the home and the woman takes care of the home and family.
                1. Strongly agree
                2. Agree
                3. Disagree
                4. Strongly disagree

FEPOL*   75. Tell me if you agree or disagree with this statement: most men are better suited emotionally for politics than are most women.
    1. Agree
    2. Disagree

HELPNOT   76. Some people feel that the government in Washington is trying to do too many things that should be left to individuals and private businesses. Others disagree and think that the government should do even more to solve our country's problems. Where would you place yourself on this scale or haven't you made up your mind on this? (Five-point scale collapsed to three categories).
    1. Government should do more
    2. Agree with both
    3. Government does too much

LETDIE1   77. When a person has a disease that cannot be cured, do you think doctors should be allowed by law to end the patient's life by some painless means if the patient and his family request it?
    1. Yes
    2. No

SUICIDE1*   78. Do you think that a person has the right to end his or her life if this person has an incurable disease?
    1. Yes
    2. No

SUICIDE4*   79. Do you think that a person has the right to end his or her life if this person is tired of living and ready to die?
    1. Yes
    2. No

LIFE   80. In general, do you find life pretty exciting, routine, or dull?
    1. Exciting
    2. Routine
    3. Dull

SPANKING

81. Do you strongly agree, agree, disagree, or strongly disagree that it is sometimes necessary to discipline a child with a good, hard spanking?
    1. Strongly agree
    2. Agree
    3. Disagree
    4. Strongly disagree

USWARY

82. Do you expect the United States to fight in another world war within the next ten years?
    1. Yes
    2. No

TEENSEX

83. What if they are in their early teens, say 14 to 16 years old? In that case, do you think sex relations before marriage are always wrong, almost always wrong, wrong only sometimes, or not wrong at all?
    1. Always wrong
    2. Almost always wrong
    3. Wrong only sometimes
    4. Not wrong at all

PILLOK

84. Do you strongly agree, agree, disagree, or strongly disagree that methods of birth control should be available to teenagers between the ages of 14 and 16 if their parents do not approve?
    1. Strongly agree
    2. Agree
    3. Disagree
    4. Strongly disagree

PORNLAW

85. Which of these statements comes closest to your feelings about pornography laws?
    1. There should be laws against the distribution of pornography whatever the age.
    2. There should be laws against the distribution of pornography to persons under 18.
    3. There should be no laws forbidding the distribution of pornography.

PRAYER

86. The United States Supreme Court has ruled that no state or local government may require the reading of the Lord's Prayer or Bible verses in public schools. What are your views on this–do you approve or disapprove of the court ruling?
    1. Approve
    2. Disapprove

RACLIVE

87. Are there any (Negroes/Blacks/African-Americans) living in this neighborhood now?
    1. Yes
    2. No

TAX

88. Do you consider the amount of federal income tax which you have to pay as too high, about right, or too low?
    1. Too high
    2. About right
    3. Too low

TAXMID

89. Generally, how would you describe taxes in America today... We mean all taxes together, including social security, income tax, sales tax, and all the rest. For those with middle incomes, are taxes . . . ("much too high" recoded with "too high," "much too low" recoded with "too low")
    1. Too high
    2. About right
    3. Too low

TAXPOOR

90. Generally, how would you describe taxes in America today... We mean all taxes together, including social security, income tax, sales tax, and all the rest. For those with low incomes, are taxes . . .("much too high" recoded with "too high," "much too low" recoded with "too low")
    1. Too high
    2. About right
    3. Too low

TAXRICH   91. Generally, how would you describe taxes in America today... We mean all taxes together, including social security, income tax, sales tax, and all the rest. For those with high incomes, are taxes . . .("much too high" recoded with "too high," "much too low" recoded with "too low")
1. Too high
2. About right
3. Too low

COLATH   92. There are always some people whose ideas are considered bad or dangerous by other people. For instance, somebody who is against all churches and religion. Should such a person be allowed to teach in a college or university, or not?
1. Yes, allowed to speak
2. Not allowed

COLCOM   93. Now, I should like to ask you some questions about a man who admits he is a Communist. Suppose he is teaching in a college. Should he be fired, or not?
1. Yes, fired
2. Not fired

COLHOMO   94. And what about a man who admits that he is a homosexual? Should such a person be allowed to teach in a college or university, or not?
1. Yes, allowed to teach
2. Not allowed

COLMIL   95. Consider a person who advocates doing away with elections and letting the military run the country. Should such a person be allowed to teach in a college or university, or not?
1. Yes, allowed to teach
2. Not allowed

COLRAC

96. Or consider a person who believes that Blacks are genetically inferior. Should such a person be allowed to teach in a college or university, or not?
    1. Yes, allowed to teach
    2. Not allowed

COURTS

97. In general, do you think the courts in this area deal too harshly or not harshly enough with criminals?
    1. Too harshly
    2. Not harshly enough
    3. About right (volunteered)

EQWLTH

98. Some people think that the government in Washington ought to reduce the income differences between the rich and the poor, perhaps by raising the taxes of wealthy families or by giving income assistance to the poor. Others think that the government should not concern itself with reducing this income difference between the rich and the poor.

    Here is a card with a scale from 1 to 7. Think of a score of 1 as meaning that the government ought to reduce the income differences between rich and poor, and a score of 7 meaning that the government should not concern itself with reducing income differences. What score between 1 and 7 comes closest to the way you feel? (CIRCLE ONE):
    1. Government should do something to reduce the income differences between the rich and poor
    2.
    3.
    4.
    5.
    6.
    7. Government should not concern itself with income differences

FAIR   99. Do you think most people would try to take advantage of
                you if they got a chance, or would they try to be fair?
                1. Would take advantage of you
                2. Would try to be fair
                3. Depends (volunteered)

HAPMAR   100. Taking things all together, how would you describe your
                marriage? Would you say that your marriage is very
                happy, pretty happy, or not too happy?
                1. Very happy
                2. Pretty happy
                3. Not too happy

Appendix 2

Answers to Matching, True-False, and Review Questions

Number in parentheses is the page number where answer is found.

Matching	True-False Questions	Review Questions	
1. 22 (4)	1. T (4)	1. d (4)	11. a (13)
2. 24 (21)	2. T (7)	2. d (4)	12. b (15)
3. 4 (6)	3. F (8)	3. b (10)	13. e (15)
4. 25 (6)	4. F (11)	4. d (5)	14. b (18)
5. 35 (7)	5. F (13)	5. c (6)	15. c (26)
6. 36 (11)	6. F (15)	6. b (6)	16. c (8)
7. 40 (14)	7. T (21)	7. d (8)	17. a (7)
8. 33 (15)	8. T (22)	8. a (7)	18. a (23)
9. 1 (13)	9. F (21)	9. d (22)	19. d (15)
10. 19 (18)	10. T (25)	10. e (11)	20. a (13)

CHAPTER 2

Matching	*True-False Questions*	*Review Questions*	
1. 19 (46)	1. T (33)	1. d (39)	11. b (38)
2. 12 (56)	2. F (39)	2. a (46)	12. e (54)
3. 24 (36)	3. F (34)	3. a (46)	13. b (56)
4. 8 (39)	4. T (37)	4. d (34)	14. a (37)
5. 6 (37)	5. F (46)	5. b (56)	15. b (36)
6. 5 (53)	6. T (46)	6. a (46)	16. c (34)
7. 20 (32)	7. F (53)	7. c (34)	17. b (36)
8. 23 (38)	8. T (56)	8. e (36)	18. a (46)
9. 4 (40)	9. T (58)	9. a (53)	19. d (46)
10. 3 (36)	10. F (56)	10. c (56)	20. d (32)

CHAPTER 3

Matching	*True-False Questions*	*Review Questions*	
1. 9 (64)	1. F (64)	1. e 64	11. b 78
2. 1 (67)	2. T (66)	2. e 64	12. d 78
3. 3 (67)	3. F (67)	3. b 64	13. d 72
4. 2 (75)	4. T 70	4. d 64	14. b 81
5. 8 (78)	5. T 71	5. b 67	15. b 78
6. 7 (72)	6. F 75	6. a 76	16. a 67
7. 4 (70)	7. F 64	7. c 75	17. c 66
8. 5 (70)	8. F 71	8. c 75	18. d 67
9. 6 (66)	9. T 64	9. a 72	19. b 70
	10. F 76	10. a 77	20. 72

CHAPTER 4

Matching	True-False Questions	Review Questions	
1. 17 105	1. F 92	1. c 92	11. d 107
2. 8 94	2. T 95	2. c 94	12. e 109
3. 26 98	3. F 97	3. c 92	13. c 111
4. 6 104	4. T 100	4. a 94	14. b 119
5. 9 92	5. T 103	5. d 93	15. d 94
6. 4 106	6. F 104	6. d 104	16. b 95
7. 1 108	7. F 106	7. a 100	17. a 107
8. 16 109	8. T 111	8. b 100	18. d 111
9. 20 105	9. F 103	9. d 103	19. d 109
10. 25 118	10. F 97	10. e 106	20. e 103

CHAPTER 5

Matching	True-False Questions	Review Questions	
1. 17 125	1. T 127	1. a 126	11. e 144
2. 4 130	2. F 131	2. b 131	12. c 145
3. 13 131	3. T 138	3. e 131	13. c 150
4. 19 134	4. F 143	4. a 130	14. c 153
5. 21 134	5. T 143	5. d 133	15. c 153
6. 27 150	6. T 151	6. e 129	16. a 156
7. 30 153	7. F 152	7. a 140	17. d 154
8. 28 152	8. F 154	8. b 143	18. b 155
9. 5 128	9. F 150	9. a 139	19. c 130
10. 8 154	10. T 145	10. d 142	20. a 143

CHAPTER 6

Matching	*True-False Questions*	*Review Questions*	
1. 8 162	1. F 162	1. d 162	11. d 178
2. 15 162	2. F 165	2. b 162	12. b 181
3. 6 164	3. T 173	3. c 162	13. e 184
4. 20 164	4. F 177	4. b 164	14. c 184
5. 10 173	5. T 179	5. e 167	15. b 171
6. 5 174	6. T 180	6. c 168	16. b 173
7. 11 179	7. T 183	7. a 170	17. b 183
8. 19 183	8. F 183	8. c 173	18. c 174
9. 3 183	9. T 170	9. e 179	19. b 161
10. 17 180	10. T 164	10. a 180	20. b 162

CHAPTER 7

Matching	*True-False Questions*	*Review Questions*	
1. 7 198	1. F 193	1. d 199	11. a 206
2. 13 199	2. T 194	2. b 190	12. a 211
3. 29 193	3. T 198	3. d 194	13. d 214
4. 30 202	4. T 199	4. b 193	14. c 214
5. 28 211	5. F 202	5. c 196	15. a 218
6. 33 211	6. F 204	6. b 199	16. e 222
7. 31 214	7. T 208	7. c 208	17. e 221
8. 1 192	8. F 214	8. a 210	18. a 204
9. 6 199	9. F 193	9. d 204	19. d 214
10. 9 195	10. T 222	10. e 205	20. c 193

CHAPTER 8

Matching	True-False Questions	Review Questions	
1. 5 233	1. F 233	1. e 231	11. b 240
2. 1 233	2. F 235	2. c 232	12. d 242
3. 25 236	3. F 236	3. a 232	13. b 242
4. 23 238	4. T 240	4. c 232	14. d 243
5. 4 234	5. F 240	5. a 244	15. a 247
6. 14 240	6. T 242	6. c 234	16. a 250
7. 16 240	7. F 243	7. c 234	17. a 240
8. 11 240	8. F 247	8. c 236	18. c 241
9. 8 242	9. T 241	9. b 239	19. b 249
10. 13 241	10. F 241	10. e 241	20. e 249

CHAPTER 9

Matching	True-False Questions	Review Questions	
1. 19 261	1. T 257	1. d 256	11. c 278
2. 14 272	2. F 257	2. a 257	12. d 275
3. 5 288	3. F 260	3. e 258	13. a 278
4. 16 271	4. F 262	4. c 250	14. d 281
5. 6 257	5. F 263	5. a 261	15. c 288
6. 20 278	6. F 271	6. a 262	16. c 277
7. 11 277	7. F 277	7. a 263	17. a 266
8. 17 288	8. F 287	8. a 264	18. b 272
9. 2 256	9. T 279	9. d 283	19. d 277
10. 3 281	10. T 275	10. d 271	20. d 292

CHAPTER 10

Matching	*True-False Questions*	*Review Questions*	
1. 5 306	1. T 300	1. b 296	11. a 311
2. 9 303	2. F 301	2. b 304	12. a 313
3. 7 307	3. T 301	3. e 306	13. b 328
4. 1 309	4. F 303	4. d 307	14. c 314
5. 4 310	5. F 307	5. b 297	15. e 326
6. 8 311	6. T 310	6. a 299	16. a 304
7. 12 313	7. F 313	7. e 316	17. b 310
8. 6 322	8. T 327	8. d 318	18. e 297
9. 2 314	9. F 314	9. c 324	19. d 297
10. 13 300	10. T 326	10. c 297	20. a 322

CHAPTER 11

Matching	*True-False Questions*	*Review Questions*	
1. 5 333	1. T 333	1. b 333	11. e 347
2. 10 338	2. F 334	2. e 334	12. d 348
3. 9 338	3. F 337	3. e 337	13. a 349
4. 7 347	4. T 338	4. c 338	14. b 357
5. 6 356	5. T 334	5. e 338	15. b 359
6. 12 332	6. T 345	6. e 344	16. c 357
7. 13 357	7. F 349	7. d 345	17. a 357
8. 8 357	8. F 357	8. a 345	18. e 338
9. 3 338	9. F 350	9. b 347	19. b 338
10. 4 350	10. T 359	10. c 347	20. c 334

CHAPTER 12

Matching	*True-False Questions*	*Review Questions*	
1. 12 371	1. F 364	1. e 370	11. c 365
2. 10 367	2. F 365	2. b 364	12. a 377
3. 9 372	3. T 367	3. c 364	13. d 386
4. 14 384	4. F 369	4. d 368	14. e 385
5. 16 379	5. F 369	5. b 367	15. d 368
6. 4 363	6. T 371	6. e 368	16. c 371
7. 1 364	7. T 377	7. c 369	17. e 376
8. 5 368	8. F 386	8. d 371	18. b 377
9. 2 385	9. T 369	9. c 371	19. d 379
10. 15 371	10. T 368	10. b 372	20. d 384

CHAPTER 13

Matching	*True-False Questions*	*Review Questions*	
1. 17 395	1. T 395	1. d 395	11. e 394
2. 9 396	2. T 396	2. a 396	12. b 395
3. 4 400	3. F 397	3. e 399	13. c 402
4. 16 405	4. T 399	4. e 397	14. c 395
5. 1 402	5. F 401	5. a 400	15. a 395
6. 5 405	6. T 404	6. b 401	16. b 395
7. 13 394	7. F 405	7. c 405	17. e 396
8. 2 395	8. F 413	8. b 402	18. d 400
9. 14 402	9. T 402	9. e 405	19. a 397
10. 11 401	10. T 406	10. d 394	20. e 418

CHAPTER 14

Matching	*True-False Questions*	*Review Questions*	
1. 18 441	1. F 423	1. c 426	11. d 440
2. 17 429	2. F 425	2. a 427	12. e 440
3. 15 429	3. F 429	3. e 428	13. c 441
4. 16 429	4. T 431	4. e 422	14. a 442
5. 22 431	5. F 431	5. d 429	15. d 442
6. 11 427	6. T 432	6. e 432	16. a 440
7. 12 440	7. T 438	7. d 433	17. d 435
8. 8 440	8. T 442	8. b 426	18. b 434
9. 4 425	9. T 432	9. c 433	19. c 429
10. 10 431	10. T 441	10. b 436	20. e 423

CHAPTER 15

Matching	*True-False Questions*	*Review Questions*	
1. 3 449	1. T 455	1. b 452	11. b 455
2. 14 453	2. F 457	2. d 455	12. e 460
3. 1 454	3. F 449	3. a 455	13. e 462
4. 7 454	4. T 454	4. c 457	14. d 460
5. 10 455	5. T 457	5. a 455	15. c 449
6. 6 457	6. F 460	6. c 457	16. e 454
7. 4 456	7. T 463	7. e 454	17. e 461
8. 11 457	8. F 453	8. d 457	18. a 449
9. 2 461	9. T 462	9. c 457	19. c 457
10. 9 454	10. T 457	10. d 455	20. d 454

CHAPTER 16

Matching	True-False Questions	Review Questions	
1. 8 467	1. F 485	1. a 467	11. b 491
2. 35 474	2. F 483	2. c 478	12. a 478
3. 25 468	3. T 491	3. d 469	13. a 483
4. 31 469	4. T 488	4. a 470	14. a 485
5. 15 470	5. T 472	5. e 473	15. b 487
6. 28 488	6. F 469	6. c 471	16. c 495
7. 1 483	7. T 470	7. c 472	17. b 471
8. 13 491	8. T 490	8. d 477	18. c 472
9. 9 495	9. T 495	9. b 469	19. e 471
10. 14 491	10. T 500	10. a 490	20. b 475

CHAPTER 17

Matching	True-False Questions	Review Questions	
1. 6 508	1. F 507	1. e 507	11. c 518
2. 8 514	2. F 514	2. d 507	12. a 518
3. 9 520	3. T 518	3. c 507	13. d 521
4. 1 507	4. T 518	4. b 508	14. b 522
5. 7 533	5. F 520	5. a 508	15. c 524
6. 10 522	6. T 522	6. b 508	16. d 523
7. 5 522	7. F 524	7. e 509	17. a 523
8. 4 524	8. F 526	8. e 510	18. c 514
9. 2 507	9. T 523	9. a 513	19. d 527
10. 3 508	10. T 528	10. d 514	20. c 527